CURTAIN *of* R

CURTAIN *of* RAIN

TEW BUNNAG

First published and distributed in 2014 by
River Books
396 Maharaj Road, Tatien, Bangkok 10200
Tel. 66 2 622-1900, 224-6686
Fax. 66 2 225-3861
E-mail: order@riverbooksbk.com
www.riverbooksbk.com

Editor: Narisa Chakrabongse
Production supervision: Paisarn Piemmettawat
Design: Ruetairat Nanta

ISBN 978 616 7339 49 8

Printed and bound in Thailand
by Bangkok Printing Co., Ltd.

Contents

Part One

Random notes from a drowning city

A screech breaks through my consciousness. Was it a baby screaming in pain, or some creature in distress? I have been lying on the bed with my eyes closed, thinking of the scene I am planning to write, floating between coherence and the usual muddle. The piercing sound makes me sit up and spill the contents of the glass I have been cradling on my belly as if it were a precious chalice. I walk over to the window and peer out. Not a sign of life, and in that grey light just before dawn the water looks black and menacing. Suddenly an enormous white bird swoops down, plucks something equally big from the dark surface and then flies away. Could it have been a sea eagle with a huge monitor lizard struggling in its claws? But sea eagles don't venture into the heart of Bangkok.

No news in the past few days. Only rumours and the occasional announcement made by a uniformed official speaking through a crackling megaphone as he is ferried speedily through the neighbourhood, promising that the water level is going down and that everything will soon return to normal. Nobody believes these reassurances, droning on without much conviction. It has been nearly two weeks, and although the top of the barbershop's swirling candy-cane sign has now begun to emerge from the murky fluid, there is a sense that the waters will be with us for many more weeks to come. I can only see that things will get worse: judging from the lurid texture of the water's surface, and the toxic stench thickening the air, an epidemic of some kind seems inevitable. Bangkok has reverted to its swampy origins: snakes and rats and giant cockroaches swirl amidst the shit and debris floating by. One morning, when the water was still at knee level, I waded through the streets to the home of the Catholic priest who lives nearby. I wanted to see if he were all right. On the way, I looked up and spotted a thick yellow python coiled in the leaves of a tall palm tree. Its tongue flicked nervously in and out of its mouth. In Ayudhya the crocodiles have begun to swim beyond the fences of the farms where they are fattened for slaughter and stretched into handbags, belts, and boots. Used to being well fed they are now hungry, and have been making their way steadily southwards into the capital to join the rest of us. They have tasted a few bites of human flesh, but no one so far has provided a full meal. The government advises anyone who comes across them to stay calm – the advice they give for just about everything else that is happening.

We are cut off from cyberspace, although I do not understand why. Since last week the electricity has been intermittent. Without warning the old fridge will rattle back to life only to die again within hours. At night there is no power at all. It is to do with the grids, we are told, and such mishaps should be expected under these stressful conditions. There seems to be a total breakdown of the infrastructure. Many people have moved to parts of Bangkok where the flooding is less severe and others, out of the city altogether. Those still here either have no choice or are afraid their homes may be plundered. Gangs are now making the rounds, looting, foraging everything they can get their hands on – the poor stealing from the poor.

I have chosen to stay to see what happens next and to record. A few months from now, when the streets and lanes are dry again, and signs of the flood have been scrubbed clean, people will forget. Our collective memory is fickle, and we are primed to be resigned. A marketplace flattened by bulldozers. Old waterways disappearing under tarmac. Noble trees that once served as shrines sliced down. Destruction in the blink of an eye. And all we do is chant the refrain we have learned: everything is impermanent, there is nothing to hold on to.

Out of laziness I was not going to bother stocking up on essentials. When the girl behind the counter at the 7-11 so cheerfully informed me that they had enough stocks to last a year I decided there was no rush to take precautions. But a few days before the dams north of Bangkok burst, the neighbours next door, Mae Lien and Nai Pot insisted on dragging me out with them one afternoon, just before the long queues started and all the shops closed for lack of supplies after the last-minute panic buying that ensued. They made me purchase a fresh gas bottle, a sack of rice and enough tinned food, matches, candles and water to last several weeks. We piled our provisions into the boot of their truck, and heaved the bags up the stairs when we got back. That day the atmosphere in the neighbourhood was still festive: sloshing water fights and lots of laughter, young boys using discarded planks from the container yard as surfboards. The fun and games stopped by the time the water reached waist height and the stray dogs and cats began hunting for a dry spot. I am grateful for Mae Lien and her son's concern and their foresight. It has been my survival – that and my adequate supply of alcohol, which I began rationing as soon as the streets turned into torrents.

It continues to rain constantly, sometimes in the form of an almost imperceptible mist, sometimes with a noisy display when the whole sky lights up and cracks open, as if the final deluge is about to arrive. The subsequent downpour produces a smacking, deafening drumbeat. Afterwards the bullfrogs' cacophonous echoes signal that the storm has passed. This is especially comforting. There have been days, too, when the sun's rays pierce the thick clouds and touch the rain, bending a complete rainbow over the city – a sight that still manages to fill me with awe.

When the phones were still working a journalist from the Bangkok News *Bangkok News* rang to ask me for a comment on the crisis. He reminded me that I had written a prophetically similar story four years earlier.

'How did you know?' he kept insisting. I tried to

tell him that it was just my imagination. This was how writers made things up. Besides, it seemed obvious that it would happen one day. He wasn't satisfied with my answer and I soon realised that he was trying to impose upon me the powers of prediction. I hung up on him.

An eeriness descends as night approaches. There's not a soul in sight these days, when once the streets would bustle with vendors wheeling their stalls from the dark narrow lanes into the main thoroughfares. Now, instead of garish neon lights the gently flickering candles glow through our windows, softening the darkness but lighting it just long enough for a meal to be prepared. It is surprising how quickly we adapt ourselves to the rhythms imposed by circumstance, when once we took things for granted. The layer of noise that was an integral part of the city has vanished – the police sirens, the thumping music, the honking horns, the cranking of machinery, the shouts and screams of arguments – and there is a deep, humming, unfamiliar silence that is occasionally broken by a dog's starving howl: a forlorn lament floating over the rooftops.

I moved here to this neglected area of the city when my mother died. My old family house belonged to another era and was burdened by too many memories. I needed distance from the physical reference points of the past in order to write about it, to recall without the presence of people I'd known most of my life. I was also licking my wounds after a failed love affair; banal betrayal, damaged ego. Writing was the only cure, and I thought I would find more time to work if I was away from the social distractions of the areas I knew well and the routine into which I had sunk in order to avoid the blank page. This isolation played to my advantage during the first couple of manic years when I spent the days and nights typing as though my sanity depended on it. It was like a penance – the need to have something to show for the privilege of not having to clock into an office or a factory every morning to do some soul destroying job, for the luxury of being an outsider. And it worked. My heart healed. The creative juices flowed. Alone and without any excuses, I felt at last that I'd found my voice and with it a way to be faithful to the Bangkok that I loved and loathed in equal measure; a city that is not just a physical location on the planet but a place of the heart – a place I had been trying to embrace since I was a young man. Then, six months ago, when the doctors revealed how ill I was, the motivation evaporated into emptiness and I sank into depression. After I heard the specialist's prognosis I stopped writing altogether. From sheer terror as to how the disease would eventually consume me I contemplated suicide. I even bought a pistol in case I had the guts to go through with it.

And then the dark clouds came and opened, falling down upon us; the rivers started to swell and overflow in the north, then in the central plains. Finally the strategies to protect Bangkok collapsed and the waters rushed into the city.

My decision not to follow the advice of friends and family and leave was not made with any courage. I wanted to see what would happen if for once I left myself with no option but to live through whatever the universe brought to my door. I was afraid like everyone else and during the first days, as the waters were rising, I thought I'd yet again made the wrong decision. But gradually I discovered that the flood, far from plunging me even deeper into my darkness, has had the opposite effect. Now, faced with the irony of living through the urban disaster I once described in fiction, I feel a surprising and renewed sense of purpose, if not enthusiasm. I have begun writing again, this time without anxiety or ambition but with a sense of lightness I have not known previously, undiscouraged by the thought that these sheets of paper could soon be floating towards the open sea along with the other detritus of the city. A new alcoholic discipline too has been established: small sips through the day and one decent long drink at night. I do not feel the urge for so much alcohol as before. Being surrounded by water and the real sense of threat and danger produces its own kind of drunkenness. Besides, I want my stock of Mekhong to last.

The waters have provided a kind of spiritual perspective. The everyday hardships I witness have made my own troubles seem almost trivial. I am no longer so scared about what will happen to my body in the months to come. I realise that if I want a quick death I can dive into these poisoned waters and drink in a couple of generous mouthfuls. That would do the job. In the meantime, I have a stash of medication to deaden the pain.

It was Mae Lien's suggestion that we eat together. She

convinced me that it was more economical to share, but I also suspect they are concerned about me. I contribute rice and a tin of food or some dried fish and she cooks it all up on her little gas stove turning simple ingredients into delicious dishes. She once worked as a cook in the household of a well-known industrialist and I joke with her that this is why she manages to feed us so well. She enjoys the compliment. I have known her and her son, Nai Pot, since I moved into this block of flats. The first day I arrived they offered to help take my stuff upstairs but I refused them offhandedly. It was a clumsy way of making it clear that I wanted to be left alone. But living in such close proximity it was impossible not to get to know one another, and we became friends within a short time. They have always made a point of seeing to it that I am OK.

Just before the waters reached the present level Nai Pot managed to get hold of a small wooden boat. I've often seen them rowing out through the lane in front of the flats taking cooked food to others who can't look after themselves. They are people whose kindness and generosity give me hope that we will come through the wreckage. Recently they have taken in a cousin called Mae Jom who was living in one of the first areas of the city to be submerged. I'd met her before, when she came from Isaan to find and later cremate her daughter. She'd stayed with them for two weeks, and Nai Pot took time from his work as a chauffeur at a luxury hotel to help her in her task. I am surprised to see her still here. I thought she had gone back to her village long ago.

We eat before the light starts to fade. During these meals they tell me the news they have heard on their radio or from the local gossip. I look forward to these daily get-togethers. Thankfully they do not mind that I have my glass of whisky, though none of them touch alcohol. Usually we part company soon after the meal is over. But sometimes one of them will start talking, recalling events from before the flooding began as if they need to keep connected to a past that is ebbing away. I have heard most of it before but it astonishes me how they manage to continue elaborating on their memories. I find increasingly that their stories wash over me and mix with my own recollections of things that happened and those that might have happened and others that have slipped over from dreamtime. It is the water's way of dissolving the boundaries between our separate realities. Carried by the same tide our experiences are beginning to merge into an amorphous expanse.

One evening I admitted to them that I have made liberal use of the stories they have been telling me since we first got to know each other, and also that in my version I have kept their real names. If they were surprised they did not show it, nor did they protest, but seemed pleased that I had done so. Naturally they are curious about what I have written and I have told them I wanted to capture the way the lives in the different strata of Bangkok are intertwined, and to chronicle the rapid changes that have taken place in our city. I didn't feel the need to explain that I'd decided to link all the stories to Nai Pot after I learned that he had once been my cousin's driver, nor that in my fiction I'd made him into my ex-lover's protector. I think this might embarrass him. But I'll get round to it before the text is published. I have promised to invite them to the book launch should it ever happen. When Clare, the editor, came from London last month just as the rains were starting, she told me they were still debating when this might be. Maybe I'll still be alive by then. But as we sit here cut off from the rest of the country, and the rest of the world for that matter, it is absurd to think that a piece of fiction has much importance. I told Clare I did not care whether the stories were published or not. She got angry at me. She would make sure they were, if it was the last thing she did, she said. This sounded ominous.

Some nights when I have broken my own rule and drunk too much, a kind of nostalgia comes over me, and I indulge in what-might-have-been, between myself and beautiful Pi Taew, who floats into my thoughts with delicious intensity. It took me years to forgive her betrayal of what we had. When I heard about her marriage, and then about her being shot, I made no effort to contact her. Now I imagine her crippled and lying in another room in the city, possibly the one where we used to meet and make love in Chinatown surrounded, as I am, by the floodwaters. I am glad that through one of my stories I found a way of giving her a kind of redemption. Perhaps when the waters have gone I will go and find her, and try to renew our friendship. I would like to meet her again beyond the limits of desire and dependence.

A seven-year-old girl drowned yesterday. Mae Lien and Nai Pot helped take her body to the local evacuation centre, which also acts as an emergency clinic. She and a friend had been using a plastic baby's cradle as a boat when it capsized. The girl panicked and in her struggle her foot became tangled in wires snarled below the surface.

When she heard about this Mae Jom was so distraught that she burst out crying: 'It's the mother's fault! If she had loved her she wouldn't have let her go out like that!' She has never been so demonstrative with her emotions. I can see how much the floods have disturbed her. She believes they are the Naga's expression of anger. She is afraid she will not be able to get back to her village where her husband is waiting for her.

The tragedy of child's drowning has had a profound effect on all of us. For several days now the meals have been shared in silence.

Our situation has forced us to return to the present and to a simplicity we had forgotten. It has certainly brought the few of us left in this building closer together. Uncertainty and death are constantly around us. I feel that through this shared plight something positive has come about and I find myself asking: can it be that the flood, terrible as it is and with all the suffering it has brought, and will bring, is a cleansing of all that has gone rotten in this city, an end to the cycle of aggressive consumerism and greed that has divided us? Can it promote the sense of community and cooperation that would make such a difference to our society? I can only pray this is true.

Mae Lien and Nai Pot have invited me to join them tomorrow morning when they deliver meals and supplies. I have said that I will, but it makes me nervous. I have spent so much time on the sidelines, observing, evaluating, re-inventing in my fiction, and in my journalism decrying the hypocrisy of our leaders and role models, and writing about how we should put our beliefs into action. Now, confronted by the reality of the situation and my own place in it. I am being invited to participate. Nai Pot tells me that I will enjoy going out on the boat and seeing this part of Bangkok with fresh eyes. I hardly think so. I am disturbed by the thought of gliding around in the black water on a small vessel not knowing what or who we will come across. But it will certainly be a new experience of the city and, besides, it is time I did something useful for other people.

I keep on thinking about what happened in that brief meeting with Clare, the editor. In the space of a few hours I felt an intimacy with her that I have never known with anyone else. It was as if we were old lovers, whose paths have crossed again. She revealed to me little about her life, yet it was not necessary to have the details. What she did share with me was the essence of who she was, the pain in her soul. We took a ride to Chinatown, to where that pain had begun, and I witnessed a moment of transformation in her. It was nothing dramatic. But I have no doubt that it was some kind of healing, which was contagious: it gave me the courage to face what lies ahead.

'Keep writing,'she told me. 'Stories are what we have left.'

She was emphatic, as though reminding me of my duty. After Clare had left I found myself quite unexpectedly thinking about a new piece, a long one this time, about an English woman like her who comes to Bangkok looking for her past.

The notes, handwritten by Tarrin Wandee finish here. There are no dates but it is a fair guess that they cover the period from late October until the middle of November 2011 when many areas of Bangkok, including the west of the river where the author was living, lay under water for over three weeks.

Chapter One

Clare spent the bright, cold Easter weekend in West Surrey with her sister Joanna and her family. On her way back to London she had looked out of the window from the train and seen a landscape that was totally unfamiliar, even though she had made the same journey many times – ever since she was a child. The fields coming to life after the hard winter, the trees beginning to bud, the rows of red brick houses on the horizon under the metallic sky; all of it meant nothing – empty reference points that left her suspended in a timeless zone without a clue to what she was doing nor where she was heading. Then, like a cork popping out of a bottle her awareness re-emerged, and she remembered that she would soon be arriving in Waterloo station. A week later she found herself trying to open her front door with her ballpoint pen. At the office she saw a name card on her desk that said: Clare Stone, Editor, Spitalfields Books, and she wondered what it referred to. The incidents were beginning to grow more frequent. When she realised what was happening she feared the worst.

'Is there any history of Alzheimer in the family?'

Clare hesitated to answer, not because she could not recall but because it was charged with pain.

'Yes,' she said finally, with a slight tinge of shame. 'My father had it.'

'Ah. At what age?'

'Sixty two,' she replied. *Too young*.

'About the same as you are now...'

The doctor's expression gave him away.

Yes, I know, she thought. *It's called an early onset. I was there with him. I lived with his panic and his courage*.

She had seen the pity on the faces of his friends, and their relatives, who eventually stopped calling by. It was too disturbing, the hearty extrovert losing his bearings, losing their names, losing himself. And now, here she was at the doorway of this journey.

'Whatever happens, life goes on,' he tried cheerily to assure her at the end of their meeting, as he was shaking her hand.

For a moment she'd wanted to slap him for such an offensively banal remark, and tell him: 'Don't patronize me.' But she controlled herself.

'Yes, I suppose it does,' she said.

The GP had referred her to a specialist in King's College Hospital, where they gave her a physical examination before doing a blood test to determine that her memory loss was not due to vitamin deficiency, or to a prior injury. She'd also been given the latest mental testing procedure called the Mini Mental State Examination, MMSE, to measure mild cognitive impairment. They made her draw pictures, recognise shapes, construct sentences, complete memory tasks. That day she was on form. Her blood was normal and none of the tests had revealed anything.

The consultant told her it was not conclusive.

'We have to see what your brain looks like. We need to do an MRI on you to be able to really tell,' he'd said, and she had agreed but when she was back home that evening she regretted not having put up any resistance.

I know what I have, she said to her reflection in the bathroom mirror as she brushed her long grey hair. *I don't need to see a picture of my brain. What I don't know is how fast it will be.*

Chapter Two

London, January 2005.

It is two weeks after the Asian tsunami and the images are still being repeated in every edition of the news. One in particular produces an intense sadness in Clare even though she has already seen it several times: a group of foreigners on a beach in South Thailand, a little girl among them pointing towards the incoming wave, which, a second or two later overwhelms them all before continuing inland to wreak it's destruction on the trees and buildings on the coast. The gasp of the witness behind the video camera is barely audible.

She turns to the two other people sitting at the table with her to catch their reaction and sees them not looking at the television on the wall but at each other, with laughing eyes. Clare knows she has just intruded on a moment of intimacy.

'Awful isn't it?' says Paul shifting in his seat and turning his attention pointedly back towards the screen. 'They still don't know how many Brits are missing.'

He uses the control to turn the TV off and then taps the polished oak table with both his palms. It is a gesture

Clare has come to recognise over the years that they have worked together as meaning that it is time to get down to business.

'So, Clare. What about that Thai writer?' he says. 'Do you think you can get him to do anything for us?'

Paul is the publisher, Clare the editorial director and Mira her assistant – all three employees of the boutique publisher known as Spitalfields Books. They are in the conference room on the top floor, rarely used by the directors. Outside the window Clare can see the bare branches of the horse chestnut shifting slightly in the wind.

'I heard him being interviewed on Radio 4 the other day. He was saying the tsunami was a wakeup call for Thailand. Spoke quite good English. He sounded very political.' Even though the question was not addressed to her Mira does not hesitate to offer her comment.

'Yes, I heard it too,' says Paul with a complicit smile.

'I'll try, if I can get hold of him,' says Clare now, slightly annoyed at being interrupted.

'Oh, I'm sure you can,' says Paul. 'After all, you gave him a leg up, didn't you?'

'A what?!' asks Mira, opening her eyes wide in mock incredulity.

Clare left the office soon afterwards, tired of the flirtatious innuendos between them. She herself had interviewed Mira a few months earlier and now she regretted recommending that they take her on, not because Mira was incompetent but because she was so blatantly ambitious. And it worried her to see how easily Paul was slipping into the kind of clichéd mess reserved for married menopausal men of his age. But she did not want to rush to judgment. She and Paul had been colleagues since she'd joined the company twenty years ago, and despite her occasional irritation towards him, they shared a casual, respectful friendship, uncomplicated by emotional entanglement.

The Thai writer Paul had referred to was Tarrin Wandee. A year previously Clare had edited a piece he had submitted to them for an anthology of writing from the Far East. The directors had recently decided to turn their attention to Asia, and they had given Paul and Clare the green light to commission whom they pleased, and thought best. A journalist from the *Guardian* had mentioned Tarrin in glowing terms and gave Clare his email address. A week later she received a polite reply as well as his manuscript.

That night she pulled out the memory stick from her handbag. She had saved his piece, 'Achaan Speaks the Truth', to remind herself why his writing had affected her.

Achaan Speaks The Truth

By Tarrin Wandee

MY MOTHER often said the mark of maturity was when you stopped blaming other people for your problems – by which she really meant parents, i.e. herself. By her criteria, I suppose, at thirty-six I could finally claim to be mature. I was in a job I hated, putting off what I really wanted to do: write. I certainly knew that my affair with a married woman who was never going to leave her husband and two small children was seriously damaging my self-esteem. And that on top of this I was steadily sliding down the slippery slope into alcoholism. I saw none of these things as anybody else's problem but my own.

When I returned to Thailand in the late nineties the country was in a bigger mess than usual as a result of the Asian economic meltdown. I had been away for nearly four years and had grown used to the well-ordered streets of Melbourne. I did not want to leave Australia and return to the chaotic city of my birth, but my scholarship had run out with my degree still unfinished and my prospects anything but bright. My father was ill; a fact my mother kept emphasizing. They were impatient for me to be back in Thailand, making my contribution to Thai society which in their eyes was about to be radically transformed yet again, this time due to the economic situation. They had been part of the student movement in the mid seventies and were still clinging on to whatever ideals they had managed to salvage from the wreckage of that brief blossoming of democracy. I had been brought up understanding that I was expected to carry on where they had left off. Now was my chance to do so, they'd said.

I returned reluctantly and spent the first months living at home, helping to care for my father, avoiding my mother's attempts to help me find a job and feeling guilty that I was not able to feed myself. I applied myself sincerely to reviving the shaky idealism of my youth, managing to convince myself that by raising the collective consciousness, it might be possible to foster the kinds of changes crucial to Thailand's future, the kind of changes my parents seemed desperate to see before they died.To that end, I took up journalism, and began to write about the national crisis and our social inequities. I was not ambitious. Saving Thailand was not on my agenda. I just wanted to make a decent contribution. But I did not anticipate the indifference I would encounter. Nor did I realise how difficult it was to live in a city where I felt like an actor playing a role that was written for me, not one that I had chosen, or could choose.

If I compared my existential anxieties with the daily challenges of others who were less privileged – those who lived in shacks overrun with rats and cockroaches and flooded with sewage in the rains, who were beseiged by loan sharks baying at the door, threatening to take the children as collateral, who were wracked by illness and without means to pay for a doctor, who were addicted to drugs or cheap liquor – I couldn't fairly say that I had any problems worth mentioning. I saw this underbelly up close when I walked the dark lanes winding through the stench of the city's slums. In my effort to bring these stories to the idle tourists and to the fenced compounds of the affluent I had witnessed the endurance and the despair.

I knew I was pathetic to even begin to complain, given the privileges I had. But I took comfort in knowing that I was not alone: Bangkok was well populated by men and women my age and with similar backgrounds who were stuck in a meaningless merry-go-round, afraid to break the mould. Most settled into a routine of unjustifiable ennui in return for not questioning too deeply what was going on around them. I met them at parties, at art galleries, in cinema lobbies, in restaurants and cafes and supermarkets. In their eyes I could see a reflection of my own dull, watery resignation, as though only a part of them were present, as though at some point they had concluded that the price of living a more creative, meaningful, challenging existence involved too much effort. I could not claim to be guided by any higher principle, but took some small solace in the idea that I questioned and doubted our collective purpose, or lack of one. This set me apart.

'Tarrin,' people I hardly knew would say, 'you think too much.'

Sometimes the words were spoken with compassion, as if I were cursed with an incurable affliction; mostly they came as an accusation, as though thinking were a contagious disease.

I never bothered to answer them that you can never think too much, or stop wondering how you are going to pass the rest of the time you have on earth. I never admitted that since my late teens I'd thought how tragic it would be to wake up at the age of sixty, the fifth cycle of life, and not feel that you had done anything except conforming to what was expected – work, marriage, children, without having taken the risk of seeing if you could have been more authentic to yourself. I was thirty-six, at the end of the third cycle, and could see that I was heading for that very tragedy.

Rescuers can be the most unlikely of people and for me it was the late Achaan Vira, a so-called pillar of the establishment, and high-profile representative of all that I so loathed in Thai society. I am sure that, were he alive, he would deny giving me any form of assistance. For someone who seemed intent on belittling me whenever he could, and undermining what little belief I had left that good would ultimately triumph, it would be galling for him that think he had done anything positive for me.

Achaan Vira, or simply Achaan as he was known throughout the nation, had been many things in his lifetime – a monk, a lawyer, a businessman, a social critic – and in his last incarnation a radio talk show host. In his fifties, he had both worldly and spiritual experience. He had famously survived a corruption scandal in which he was clearly culpable, emerging with his reputation unscathed. It fascinates me that someone like him, with so few praiseworthy attributes other than a cunning intelligence and the thick-skinned determination of a survivor, managed to become, at least for a while, an arbiter whose opinions influenced Thai society's behaviour, if not their thoughts.

He was born in the Thonburi district of Bangkok in 1952 and was promptly abandoned on the steps of a temple. The monks took him in, clothed, fed and educated him. He became a novice and was eventually ordained. He was clever and hard-working, so the abbot sent him to study at the Buddhist University in Ayudhya, where, upon graduation, he shed his robes and found work as a clerk in a rice-exporting company. He rose quickly to assistant manager while studying law at night. When he obtained his degree he joined a law firm specializing in surveying, and land contracts.

All this is a matter of the public record, and there are many who claim friendship with Achaan from these years willing to substantiate the details of his early life, after which the waters grow more turbid. In any case, by his late thirties he was a millionaire several times over. There is considerable speculation about how he managed to scale such heights in so short a time, but in the absence of evidence to the contrary, the 'official' line – that he was a wise and skilful investor – is difficult to challenge. Certainly, anyone ruthless and astute enough in the early 1980s in Bangkok could make an enormous profit on urban development in the housing market. And that is as it appears on paper.

Rumour and gossip told a different story. Even my father, a man not prone to idle chatter, would look up from the newspaper and say to my mother: 'One day that man will go too far; he's a high class pimp and he's got everyone eating out of the palm of his hand.'

Bangkok was engorged with the self-indulgence of newly minted millionaires like Achaan and the excesses triggered a well-organized backlash among conservative Buddhists. During an investigation into Achaan's finances in 1987 it came to light that he had been involved the 'entertainment' business since the mid-1970s; that is, the growing sex trade that appealed to a particular type of European and Asian tourist. Achaan, it transpired, was a major shareholder in Thaifun Co. Ltd., which ran two of Bangkok's most successful upmarket 'massage' establishments and more than a few of Bangkok's most glamorous nightclubs. During a police raid on a warehouse in the port district of Klongtoey, a heroin dealer was arrested and grabbed the headlines with his revelations that he supplied many Thai stars and starlets – he produced as evidence photographs of himself in their company, which the newspapers lapped up – and alleged that Thaifun was a front for drug distribution and trafficking. More pictures again, this time of him with its corporate officers and other major business figures. Against all the evidence Achaan Vira and his associates maintained their innocence, their lawyers contending variously that the man had duped them by pretending to be a potential investor, that he had been refused employment because of his drug connections, and that he harboured a grudge against the company. Once the scandal had quietened the dealer recanted. He was given a suspended sentence for creating public mischief and was never heard from again. Though no charges were ever laid against Thaifun or any individuals, Achaan's reputation was temporarily tarnished, the self-styled guardians of the nation's morality citing him as a contributor to Thailand's sleazy international image as a destination for sex and drugs and rock and roll.

In reaction to the international bad press Thailand in the early 1990s expended a lot of energy reinventing itself as a wholesome tourist destination and ultimately it was in everyone's interests that the scandal be forgotten. Achaan did his bit by retreating to the temple that had shielded him as a boy and, during one rainy season, he took again the vows and robes of a monk. Publicly, it was seen as an act of atonement. When he emerged from his meditations, a layman spiritually renewed, he issued a statement in which he admitted 'past mistakes' and vowed he would have nothing to do with any business that might tarnish Thailand's reputation. And with that, he began campaigning against the sex trade with the zeal of a missionary and, through a weekly column 'My

City' in the popular Thai-language tabloid *Daily Bangkok*, he thundered fanatically against the 'entertainment industry' and, when the public began to tire of the subject, he took on other 'immoral' aspects of Thai society: manners, clothing, hairstyles, films, language. Nothing escaped his scrutiny and disapproval; he maintained that what the country needed was a return to traditional Thai values. The tone of his delivery, and his message, caught the mood of the times, reflecting as it did the ideas of the country's increasingly conservative leadership and those who felt powerless in the face of rapid and aggressive change. 'Thailand is in danger ...' was a popular Achaan refrain: in danger of losing its core values; in danger of losing its identity; in danger of becoming a tourist playground where the Thai people existed only for the pleasure of others. Achaan had tapped a vein of dissatisfaction. None of his words, and none of the well-publicized campaigns by those who had the authority to clean up Bangkok, made the slightest difference to the sex and drugs trade, both of which grew steadily and fruitfully. Achaan was undeterred. He became a regular commentator on a television current affairs programme, he opened shopping malls, he was photographed holding babies in his arms. He now wielded considerable power as a political observer and politicians who once rushed to distance themselves from him began to court his support. However, when he received an invitation to join one of the more established parties Achaan, against the expectations of the general public, announced that while his first duty was to serve the Thai people, he wanted time to write a book on Thai history. He said he was even considering returning to life as a monk for the third time. Again the rumours were rife: that he was seriously ill, that he had AIDS; he had a drinking problem; he was a drug addict and was going abroad to detox.

The truth was more mundane. What gradually emerged was that he had lost a great part of his wealth with the collapse of a private Thai-European airline, in which he was a major shareholder. The bursting of the bubble in 1997 had been a long time coming. Everyone knew the economy of the country had been managed with a gambler's foolhardiness; everyone borrowed from anyone who would lend. Achaan was just another casualty. Despite his inflated ego he suffered a breakdown, and disappeared completely from view.

The country's recovery was slow and painful. I have no idea how painful it was for Achaan Vira because I have no knowledge of what he did during the years after the economic crash. All I know is that his second successful return to public life coincided with my disenchantment with everything that had sustained me since returning to Thailand.

To have a break from Bangkok I'd headed down to Haatyai, in the deep south where a complicated and volatile conflict between the government and the Muslim separatists was coming to a head. I had done my research and thought I had a story that would have some impact on opinion. It did not. The government under Taksin Shinawatra issued an Emergency Decree, like the one they'd used in their so-called War on Drugs. Demagoguery yet again triumphed over reason. Hundreds were killed and a new generation of angry Muslims despaired at any hope of finding a political solution. This marked for me the final acceptance that I was powerless to make any difference. My words could not even raise an eyebrow. I returned to the capital knowing that I did not want to spend the rest of my life dealing with causes bound to fail because they went counter to the vested interests of those who wielded power. In fact, I was through with trying to follow the twists and turns of Thai politics, of keeping up with the latest intrigues and rumours, of analysing situations and events in search of some hidden subtext. That unsavoury world had turned me into a cynic. The future looked bleak and I was tired and emotionally drained and convinced that the only answers lay at the bottom of a glass.

Enter Achaan Vira.

I started working with him on a radio programme that became one of the most popular and successful in the country. It was something that fell into my lap. Khun Lung Serm, a close friend of my father and recently appointed head of Radio 9, had coaxed Achaan out of retirement and persuaded him to return to the world of media. He convinced him that his skills as a communicator were needed and offered him a programme called 'Pood Kwaam Jing' – 'Telling the Truth'. It was to be a chat show offering political commentary, a satirical take on world affairs, and advice to listeners on legal or personal matters. There was nothing new about the concept; others had tried the format with varying degrees of success, but what they lacked was the personality and charisma of someone like Achaan. Playing to his ego, Khun Lung Serm couched his pitch in terms of educating the public, moulding a responsible and decent Thai citizenry. He also offered Achaan a lot of money.

Khun Lung Serm had been in the student movement with both my parents and was later a colleague of my father's at Thammasat University where he taught history before leaving the academic world to join the private sector. I remembered him as a kindly professor type who had shown interest in my studies, but I was unaware he had been following my career until he phoned to offer me the job of researcher on this new programme. My mother told me later that Khun Lung Serm was concerned about what I was writing and, out of affection for my late father, hoped to put me on a safer course before I burned too many bridges.

'It's a piece of cake, Tarrin,' he kept saying. 'With your English and journalistic background, you can do it blindfolded.'

My work, he told me, was to scour the local and foreign press for odd or eccentrically amusing items and then write summaries Achaan could use during his two hours on air. *Easy enough*, I thought, and imagined how I might best fill the coming months of paid idleness. So I agreed and met Khun Lung Serm at his brightly lit office on top of one of the newer tower blocks. It was supposed to be a friendly 'getting-to-know-you' session with Achaan Vira. Khun Lung had used the English phrase, conjuring for me that scene in the 1950s musical *The King and I* with Deborah Kerr waltzing across the palace with the wives and children of Rama IV. The film was banned in Thailand but I had seen it by chance on television in Australia. The three of us didn't dance but we did drink a great deal of Mekong whisky, cut with soda and lime – Achaan's favourite drink – and spent two hours sunk in leather armchairs looking out over the city sprawl, listening to a monologue by him on wildly digressive topics, only one of which I recall: the weaknesses of the British monarchy compared to our Thailand's rock solid and unassailable institution. 'They should have held on to lèse majesté' is a comment that sticks to my mind.

Achaan's views and political stance were 'conservative'; I already knew that from reading his column in the *Daily Bangkok* and watching television programmes on which he'd appeared. But the mental picture I had created of him was nothing like the man in real life. He was tall and dark- skinned, with a flat, broad face that gave nothing away. There were deep rings under his eyes and his immaculately cut grey hair was precisely parted on the left and slicked smoothly down. He wore an amulet suspended on a thick gold chain around his neck. A scent of expensive cologne entered the room with him. His body shape was that of a bulbous pear, one that has been injected with water to increase its bulk. His handsome printed-silk shirt was tight around his belly, and as he stretched his long legs I saw purple socks above the tops of soft, brown-leather loafers that looked as if they cost a modest month's salary. The overall effect was of someone fussily conscious of appearance. I had expected him to be difficult and critical, but I was entirely unprepared for his bombastic mix of populist sophistry that was miles to the right of anything I had come across before: racist, anti-intellectual and contemptuous of those who would put anything, especially the environment, before profit. He was possessed of a cruel sense of humour and an unshakeable belief in his own infallibility, having shrugged off somewhere along the way any pretence of being open to opinions not his own. His thought process seemed to be fuelled by alcohol and bitterness. His voice sounded reptilian, its tone growing weirder with each successive glass of whisky. Khun Lung kept whispering sotto voce praise to me of its fantastic potential on the airwaves, which Achaan feigned to not hear.

Throughout the meeting it was clear that Achaan considered me a young and grossly uninformed idiot contaminated by having spent years away from Thailand studying Asian politics.

'All that farang culture, it makes you stupid,' he pronounced, cutting off my fumbling response with what I would learn was a habitual, dismissive wave, so often delivered that it might have been tourettic. It was a rapid, impatient gesture he often made over the microphone, like some malevolent, priestly act before delivering one of his scathing responses to a remark from a phone-in listener he deemed moronic, annoying or on the brink of making sense. When Khun Lung Serm tried to promote me as an invaluable research assistant, Achaan chuckled and shrugged and, offensively stabbing the air with his forefinger in my direction, said:

'Well, at least you learned some English. Just give me what I want, not what you think I want.'

Khun Lung made no attempt to steer the meeting and by the time Achaan declared himself late for another appointment I knew I did not like what I had seen and heard.

'I'll see you on Monday, young man,' he shouted as he weaved towards the lifts. 'Don't be late. I like people I work with to be punctual.'

As the doors closed on his seething form, I had an image of the Medusa, a head of writhing snakes.

The moment the elevator light blinked off, Khun Lung spun around to face me and, with a tone of urgency that scarcely left time for breath, said: 'Do not judge him, Tarrin. Do not be so critical. You've seen only a part of the whole. Give him a chance. Achaan's got something very special. He'll draw them in. He'll be popular again. Trust me. I can sense it. He's a self-made man who's worked very hard in his life. He's pulled himself all the way up to where he is now by his own effort. No family connections, no one to give him an easy ride. I'm sure you're acquainted with his background. It's fascinating. People can identify with this kind of man. So don't knock him. He may have strong opinions and his style is not always to everyone's liking but he's a good person. He's always making merit at the temple. He was once a monk, you know. You'll make a great team.'

Khun Lung's face betrayed desperation, and he added what sounded to me like a plea: 'Please don't let me down.'

The lift pinged several floors down, interrupting the cool silence induced by all the marble cladding. It had been a surreal afternoon. Obligation to Khun Lung aside, money made the decision for me: his offer was generous and I wasn't a 'rich kid', so I agreed to do as he asked, knowing full well that there would be no 'team', great or otherwise. I knew, just as Achaan knew, that ours would be a relationship of conflict, not violent and overt but ice-cold.

It took some effort to force myself back into a routine: in the office by nine-thirty, depending on the traffic, I would start by sifting through the piles of Thai and English-language newspapers stacked each morning on my desk, tearing out whole pages, branding articles with a neon yellow marker, checking online what the foreign media were saying. By noon I would have in front of me a list of news items grouped by category, my comments in the margins, ready for Achaan to come breezing through the door, me standing, palms respectfully together and bowing a wai, an act repeated by other staff and acknowledged perfunctorily by him without attempting the slightest eye contact as he headed for his desk, pulled open the bottom left drawer to extract a bottle of Mekong, which he banged down in front of him before burying his head in the clippings. This was the signal for me to ring down to the bar on the first floor for two tall glasses, a bucket of ice and bottles of soda water. I mixed two drinks – this became my job from the first day – handed one to Achaan, which he finished in one swallow, and took one myself. I sipped at mine. I then would make him another, and with his second drink at hand, he would question me about the material I had collected. His voice was flat, strictly matter-of-fact, peppered with sarcasm to let me know he thought my efforts mediocre, my choices weak, and that he held me personally responsible for the dearth of salacious material his programme consumed.

He gave me no clue as to what he wanted, nor did I ever manage to intuit how to make him happy. Thinking that he would be returning to the themes that had made him a popular columnist, I scanned the pages for articles to do with the moral decline of our culture. But he had moved on from that particular crusade and, in so far as I could divine, the moralist had turned into cynic. He did on occasion berate me for failing to keep watch for evidence affirming his belief that foreigners – that is, white farangs – were fools who had no idea how to run their own countries, let alone dictate terms to others. In his mind, Thais were simply cleverer and more skilled and altogether superior in just about everything. His was not a unique view in geographical Asia; one need only change the noun to fit the place – 'Malaysians', 'Singaporeans', 'Indonesians', 'Chinese',

'Indians', most especially 'Japanese'. I always assume that those who express such beliefs do so out of some long-standing grudge over a bad experience deep in the past with a foreigner. Europeans were the easiest to hate, courtesy of history.

Perhaps he was still bitter about the failure of his airline. Perhaps someone had stepped clumsily on his toes. It was impossible to tell. But the opportunity to vent his views gave him not only satisfaction but inspiration, because he could extract all sorts of insights and interpretations as to how, in the end, life was really much better in Thailand than anywhere else.

Thus would begin an afternoon of radio that might start off with a declaration that, unlike Westerners, we did not spend time agonizing over our motives or harking back to the past. We would not start feeling guilty like the Americans for the atrocities committed during the Vietnam War. We were free of guilt. We had no hang ups about sex although, in his considered opinion, we were perhaps a little too liberal about it. In general his tone on sexual matters was disapproving: ladyboys were as confused themselves as they confused others and he constantly made jokes about effeminate men and gays that revealed a deep-seated homophobia. But if Achaan were at all aware of his bigotry it would have been with a sense of pride rather than shame because he liked to assert a macho, devil-may-care persona that he claimed befitting to the Thai male.

Whenever I found a good example of farang moral confusion or double standards or contradiction he would congratulate me and declare me 'astute'. It didn't help much. Once, overcoming doubts as to whether he might be offended by seeing anything that could be taken to allude to his own past, I passed him my notes on a bribery scandal involving a reputable British company and the heads of states of various African countries. Achaan was over the moon.

'They call that "corruption"?' he bellowed down the microphone to the imagined laughter and approval of his audience. 'I call it "chickenshit". Bribing heads of states to get your contracts through is normal procedure; nothing wrong in that. We Thais could show the farangs a thing or two about real corruption ...' and he continued for the next half an hour without a hint of irony.

The show started just after the one o'clock news. There would be a jingle and then Achaan's recorded voice came on booming: 'Pom pood kwaam jing kraab', which roughly translates as 'I tell the truth'. Even now when I hear that phrase I shudder quietly at its hollowness. He would hit a button, initiating a digital recording of raucous cheers that filled the airwaves and off we went until three o'clock, two hours of busily idle time in Bangkok that got him an audience of drivers, office workers dozy from the heat, and tie-less business types on their way to or from some meeting. As the programme went on air, I worked with Nong Ta, the secretary, screening incoming calls for Achaan to take, either on the day's topic or something he could comment on or supposedly fix given his vast legal knowledge, wisdom, worldly experience or network of contacts. Listeners lapped up his homey insights and sense of humour. Many were faithful followers pleased to have him back as a commentator. If they noticed at all how his attitudes had hardened and become more extreme and bitter, they either didn't care or were just as extreme and bitter as he and found Achaan's xenophobic world view reassuring, almost comforting, even as he cajoled and insulted these very same people with a ruthlessness that made me cringe. They adored him. I think he relished finding fault with everyone, and made no effort to mask his contempt for those foolish enough to disagree with him. Informed debate was not on his agenda. But it made no difference to his fans.

'Achaan Vira's the only one who tells it like it is,' said a taxi driver to me one day as we sat idling in smog so acrid I could feel it on my teeth. 'The others, especially the politicians, they're all crooks out for their own profits. Achaan's been through all that. He knows the truth and he's not afraid to say it. Me and my friends, we all listen to him. Who else can we trust?'

I was stoic in my silence.

But the taxi driver was wrong; in Achaan's eyes, not every politician was corrupt and he could see no wrong in one particular populist politician with a nationalist, if not downright fascist, message – Sompop Wongpanich. When I first heard him mention the name I nearly dropped the glass of Mekong I was holding. On reflection it should not have surprised me that Achaan's admiration was focused on such an unpleasant human being and one that I knew personally. For this same Sompop was a cousin, on my mother's side and about ten years older than me. Our families were not close; Sompop's clan was in an entirely superior economic bracket, with their grandfather having made his wealth in construction. Blood ties meant we saw one another at family functions. My father avoided interaction with them, which confused me until I grew old enough to understand, and it intrigued me to watch Sompop's engineered entry into politics and his rapid rise through the ranks on the back of his earnest fears about what was happening to the country and his businesslike solutions to the problems. In our country fear is the key to the art of manipulation. A sweep of the hand is usually enough, and Sompop liked to wave his hands about. I'm not entirely sure he didn't steal the mannerism from Achaan. That we were related I kept from Achaan after due consideration. I decided that I did not want the fact to impress him in all the wrong ways, nor to spoil my perverse enjoyment of hearing him gushing with praise bordering on adoration for my distant cousin whose opinions I had always tended to ridicule. It was obvious that Achaan saw in Sompop a saviour for Thai society. For my part I saw him as a character to be stored in my notebook for future use.

When the programme started, I honestly did not expect it to last more than a few weeks. I could not imagine there were people out there who would listen to such rubbish. But midway through the third week

Khun Lung Serm was brandishing a sheaf of paper that purported to show a surge in listeners and within a few months we had won the timeslot. Khun Lung poured drinks and insisted he had never doubted Achaan's popularity, going so far as to tell me privately that I might be due a pay rise if I kept Achaan top dog. I suppose I should have been pleased, but instead found myself weighing the monetary value of hypocrisy and reasoning that as I was already a whore, I might as well be a highly paid one.

I quickly built up an immunity to Achaan's brand of rhetoric to the point where I scarcely heard what he was saying. Ask me what yesterday's topic had been was like asking me the speed of light – my brain would tell me it was something I knew but could not at that moment recall. Once the initial comfort of having a steady, paid job faded, I would wake everyday with the hope that it was all some horrible nightmare and go to bed that evening planning to do something better with my life. I blamed no one but myself. Nobody was forcing me to do anything. I had created my own karma, as the Buddhists say: I had planted the seeds of my life and my tree had borne a sour fruit.

What kept me from total despair was the easy and pleasurable affair I was having with a woman called Pi Taew, who worked for an insurance company in the tower block next door. We had met in the café on the ground floor of her building the week after I began at 'Pood Kwaam Jing'. I had arranged to meet a former colleague one afternoon for a coffee and she had bought along a friend, Pi Taew, who was trim and stylish, with a short, sporty haircut and whose knowing expression invited conversation. I found out within minutes that Pi Taew was a few years older than me and that she was married with two small children. She seemed impressed that I worked for Achaan Vira; she listened to the programme when stuck in traffic, she said, and found him 'interesting'. Her eyes were laughing when she said this, and gave a hint of collusion. I was at once relieved that she was not a fan of his views, intrigued by her skilfully diplomatic choice of words, and cheered that she had made an effort not to offend me. I doubt I would have noticed even if she had. We talked about a recent film the both of us liked, and about the book I could see protruding from her enormous and overfilled handbag. My colleague, who had gone to fetch the coffees, could see straightaway that she was now excluded from our increasingly friendly exchange and, not long afterwards, made her discreet excuses and slipped away. I sensed a rising empathy between Pi Taew and myself and at some point she accepted my invitation for her to join me for a drink the following evening. I told her, with the nervousness of a man who was about to embark on a relationship with a married woman for the first time, that I did not want to cause her any trouble, that I would never come between her and what she needed to do with her family. I remember that she'd touched my arm lightly as she reassured me that if we both acted maturely, no one would be hurt. And that was how our affair started.

Pi Taew had a 'flexible' schedule as a travelling salesperson for her company and when I'd finish work we'd have a couple of hours together three afternoons a week, four if her family commitments permitted. We'd meet in a café or noodle shop and take a taxi up to Chinatown where we went to a flat in an old, tall building in a lane near Sampeng market. The building had no lifts, only a narrow wooden staircase, and the air smelled of medicinal herbs being boiled in some back kitchen. The first time Pi Taew took me there a Chinese love song was playing in a room along the hallway and she hummed along gently to its refrain. She had 'borrowed' the flat from a friend, she told me. The windows of the sitting room were adorned with lacy curtains. The bedroom featured a ceiling mirror and the bathroom a jacuzzi where Pi Taew and I would sit and drink chilled white wine as though we had all the time in the world. Then we would make love and afterwards drink more wine and talk about literature and film, and the way the country was going, and I would make her laugh with my imitations of Achaan Vira and his grotesque ways, and she would tell me a few things about herself: about being Hakka and how her great grandfather had arrived penniless from China. Now and again, at my prompting, she would mention her husband, whom I learned was a policeman, and her children. But she would quickly change the subject. I could see she did not want to share her other life with me.

During those delicious months, which now seem so distant and dreamlike, I learned to endure Achaan's surliness and unrestrained on-air hectoring so that three times a week, four if I were lucky, I could have good sex in decadent surroundings with a beautiful woman who demanded nothing more than for me to take my pleasure with her body and engage in intelligent conversation. My initial qualms about entering into an affair with her gave way to a sense of satisfaction at how fluidly it was all working out. I had accepted long ago that I was unsuited for committed relationships. I liked being single and having my own space and was nervous of any serious involvement with anyone. That Pi Taew, my Chinese Thai lover, had a family meant I was safe. She would never leave them, and I would never ask her to do so. We were both happy.

That this happiness did not last was, I suppose, entirely my fault. I severely miscalculated the limits of my involvement, thinking I was in control of my desire and of my emotions. I found out the truth the Buddhists like to emphasize: that desire leads us by the nose and before we know it we are hooked. I did not even see it coming. Despite my intention to remain independent I grew more and more attached. As time went by I began to feel the need to see her, not every other day, but every day, and outside our prescribed hours. I yearned for the feel of her skin, the smell of her perfume, her tinkling laughter. I wanted her all the time. It was a passion I had not known before, an urge to have this woman all to myself, a kind of animal possessiveness that began to obsess me with a force beyond my control. I could not stand the thought of her in someone else's arms. I came to hate her husband, whose picture I had not even been shown but whose rights came before mine. At first she took my ardour for flattery.

'But I have a family,' she would gently remind me. She said this teasingly at first, but after a while she became openly annoyed whenever I asked if we could spend more time together, a weekend on Koh Samet island or a trip to the mountains around Chiang Mai. In my obsession I had planned things to such an extent that I could confidently say, 'Let's just go, now.' As if anticipating such a move on my part, she always had an arsenal of excuses that were hard to challenge and, anyway, she had made things clear from the beginning. Though her inaccessibility fuelled my attraction, it also pushed me towards petulance, and as she drew away, I wanted only more desperately to get closer to her. Masochism has a lucrative market in certain parts of Bangkok. I got to indulge in it for free and grew sick with jealousy, unable to stop thinking of her and dreaming of a future with her by my side.

It became so bad that when I wasn't with her, I drank, and drank heavily, mainly by myself at night but also with Achaan, who'd seem pleased when we'd drain the day's second bottle of Mekong. Drinking did nothing to soften his dislike of me, though he was not one for letting disagreement get in the way of a good monologue.

'So you want to be a writer,' he said. 'You don't have the self discipline, Tarrin, you drink too much. But forget all that. Manga type comics are the future. Lots of pictures. Violence. Action. Strong, recognizable characters. Not too many words. Not too much to think about. Makes it easier for everyone.' He waved towards the bottle, which was tracking our progress, and signalled for me to mix another round.

'You're an unhappy young man,' he told me on another occasion, while watching the fresh ice cubes melt into the liquor. 'That's obvious enough. What you need is a girlfriend, no one special, just a nice plain girl who won't mess you about. Not that they are easy to find nowadays. And you are not exactly a great catch for anyone. Or are you gay? If you're gay, be sure to protect yourself when you go to those clubs, or else you're bound to pick up some awful disease.' Then he chuckled to himself.

I looked into my glass, thanked him for his concern, and silently cursed him.

One morning Pi Taew called from her office and asked if we could meet for an early lunch in a new Italian restaurant nearby, which seemed odd because she was never usually free until mid afternoon. But there was no way I was going to refuse, even though it meant I had to make up an excuse to leave the office when the programme was on air. I told Achaan I had to visit my mother in hospital, and that it was urgent. My mother had cancer and was being treated in hospital – that part was true – but visiting hours were in the evening. Achaan didn't believe me for one moment but it would have been impolite to challenge my lie in front of staff, and to question such a personal matter, so he shrugged and with a crude gesture of his thumb motioned me out.

Pi Taew was waiting at a discreet corner table, where she sat idly playing with a packet of sugar. We drank some wine and then she told me outright that she had decided it would be better for us not to see each other any more. It was over. I had been half expecting her to tell me this for some time; lately she had seemed increasingly distracted during our assignations, and less attentive. The pain of hearing her pronouncement was no less delicious for having been anticipated. I made her explain in detail her reasons as I protested and told her how much I loved her. It was simple, she said. Her husband, a middle-ranking policeman, was up for promotion, which would mean a higher public profile. He needed a devoted wife at his side to bolster his image as a caring family man. Recently he had become suspicious and had hired a private detective to follow her. The detective wasn't very good, or wasn't being paid enough, or her husband didn't really want to know but wanted her merely to be aware that he was onto her, because until now she'd had no problem giving him the slip. But she couldn't take the tension any more and was afraid of what her husband might do to me.

'When he's angry he's mean,' she said. 'He won't do anything himself. But you know he has connections. He scares me. When people make him angry, he mutters to himself, as if I can't hear. You know what he says? He says, "Just another corpse in the morgue." The thought of you lying on some cold slab –' she broke off, and started to cry.

As I listened to her, all I heard was that her husband's pride had been hurt. He probably knew all about me and could have had me disappear at any time. With a bravado fortified by desire I told her I couldn't care less. It was beyond my meagre power to do anything about him. But as for Pi Taew, I was convinced that she loved me. Seeing how upset she was I kept telling her that I would cause no trouble, and that I would agree not to see her for a while if it made things easier. But I also told her that I was deeply in love with her and would never let her go, and that when things had calmed down and her husband had got his promotion we would see how we could carry on. She listened without interrupting. But when I'd finished she merely sighed and shook her head from side to side as though to emphasize the impossibility of our situation. In the course of the untouched meal I did manage to persuade her to spend one last afternoon with me to say a proper goodbye, and we took a taxi to Chinatown and made love and parted, as had become our habit, at five thirty on the dot.

That night I didn't drink much and was awake until dawn, going over and over in my head what she had told me. I couldn't decide whether she was telling the truth, or whether she had simply been playing with me all this time, like a shiny toy she had wanted but with which she had now grown bored, or found something shinier. So many contradictions: a letter addressed to her, at the apartment that supposedly belonged to a girlfriend; her policeman husband, or had she said he was in the army; even the children, which had left no childbirth marks on her, and I had examined every inch of her body.

Why should she have told me the truth? I was just her afternoon lover.

I watched the sunrise gradually illuminate the bedroom through the spaced slats of the shutters, my mind exhausted by speculation, and then it came to me, a kind of satori, a revelation impelled by a moment of supreme self-delusion as I remembered her tears. They were real. And she had shed them for me. Pi Taew really did want to be with me and it was my mission to save her from a marriage that had become a sham. I cast aside the warning that we must never meet again and embraced the risk that her truth implied, that I was risking my life in pursuing her. I felt calm. My mind was at rest. And I slept for maybe just an hour before the phone woke me. It was Khun Lung Serm.

'Tarrin, are you awake?' His voice was edged with urgency. 'Get over here as soon as you can. We have a serious problem.' And the line went dead.

Nong Ta was already at the office when I got there and told me Khun Serm was waiting. Her normally blank expression, which served her well as an observer of events, now betrayed the slightest hint of worry. She said nothing, but directed me towards the inner offices. I found Khun Lung Serm silhouetted with his back to me, looking out through the tall windows, hands clasped behind him. The sun had risen but its light was still soft and the blackening clouds that were gathering in the morning sky were trimmed with silver. I thought I was about to be sacked, and felt elated.

'You have to help me, Tarrin.'

Khun Lung turned to face me. His expression was heavy with worry and fatigue.

'Anything,' I said.

'An obituary,' he said. 'No ... I think a eulogy.'

'Of course,' I said, puzzled. 'For whom?'

Khun Lung Serm cleared his throat, as though trying to dislodge a fish bone.

'Achaan Vira.'

Khun Lung Serm looked at me, waiting for a response, and repeated, louder this time: 'Achaan Vira is dead.'

'But ... but, yesterday... he looked fine.'

Khun Lung Serm held up his right hand, palm towards me, shutting me up. There was a surging sound in my ears, like crashing waves. The sunlight caught the facets of the small ruby Khun Lung wore on his right pinky and I was for a moment lost in its infinite depths.

'Tarrin ... Tarrin ... listen. You were his colleague, and he had great respect for you. I don't think the programme would have been such a success without you ... But, something has happened to Achaan and you should know the details.'

He paced back to his desk and sat behind it.

'This story does not leave this room. Do you understand? Swear to me that it won't.'

I nodded.

'It seems Achaan was murdered; it was...it was...oh,

what is it the French call it, when something happens in the heat of the moment?'

'*Un crime passionel* ... a crime of passion.'

'Yes, that's it. A crime of passion. He was murdered ... late last night ... in his own house ...'

I said nothing.

'...by his gay lover ... who then killed himself.'

It was a lot to take in. My world had just turned upside down.

'His poor wife,' said Khun Lung Serm, shaking his head in sympathy.

I had met her only once, at the reception party that launched the programme. She had pale skin, was thin and small, like an elegant bird, and dressed on that occasion in traditional Thai costume. She looked as if she were from another era. She wore a sad smile as she greeted guests and when we were introduced I thought I saw a look of pity, though perhaps I was just projecting.

'... right there in the bedroom ... in front of her ... he must have had a key. He was a young man, probably a professional of some sort. There are so many of them these days. They make good money from the farangs, from the club scene. Oh, you must know all about that, your generation ... ghastly business. He was probably high on ya baa ... so many kids these days are hooked on that crappy speed ... He screamed at Achaan and insulted his wife, in their own house. Can you imagine? Then he took out a knife, stabbed Achaan several times and slit his throat from ear to ear. Blood everywhere ... must have got an artery early because it sprayed all over the place. His wife remained calm ... probably what saved her ... and after that it seems that the young man plunged the knife into his own heart. It must have been terrible for her. But she kept calm. Nerves of steel. And such a small woman. Called her brother, a high-ranking policeman, a real big shot, and asked him to deal with it. He'll make sure there is no scandal.'

I was about to ask the obvious questions but Khun Lung Serm again signalled silence.

'Everything's been taken care of. The house is all cleaned up. The servants won't be a problem and the ambulance came from the police hospital. Everybody involved has been paid or warned to keep their mouths shut. The official line is that Achaan died last night from a heart attack. Quite normal at his age, I should think.' He paused. 'The only thing left to do is to for us to close down the programme. What a shame. The ratings were so good... Anyway, I thought a decent obituary would be a good way to do it, and you're going to write it. I'll read it out on air after the one o'clock news. That way, I pay my respects to him and explain what's happened. I've already called the technician who's putting together some snippets from old shows, and we'll replay me at the bottom and top of each hour.'

Up till then I had vaguely thought that Khun Lung Serm, like my parents, had secretly retained the ideals of his youth, and that his values were based on some sense of integrity. Now I realised he had done away with the things that did not matter in our modern Thailand. He was first and foremost a businessman, and he needed to have everything under control. By all accounts the boat must not be rocked.

I, on the other hand was in total shock. My head was reeling. All the things I had not said to Khun Lung Serm when he was telling me the horrific news, and all the questions I might have asked when he had finished had I been more awake now raced through my mind but found no answers. Once I'd left Khun Lung's office I went over to Achaan's desk and opened the bottom drawer. Three bottles of Mekong, one of them perhaps a third full. I felt for a moment physically sick but the wave of nausea passed and I took a long swig, not bothering to find a glass or order down for ice and soda. The alcohol tasted like perfumed paint thinner.

Achaan's desk was a no-go area when he wasn't in residence. Not even the cleaners were allowed near it. After each programme we would often see him writing on a sheet of paper, which, upon leaving for the day, he would fold and put into his briefcase. I saw on his desk just such a sheet, and written on it in his hand was '6.30'. I presumed it had been written the day before, perhaps as he'd talked on his mobile phone, perhaps in conversation with the very man who'd later killed him. I opened the other drawers of his desk, but they were empty. What was I looking for? Clues to his other life? I thought I knew him because he seemed so uninhibited: a bigoted, self-righteous egomaniac inhabited by ugly visions. He'd made no secret of that. In death, he was a mystery, just like everyone else, possessed of shadows and demons.

When I looked up from my thoughts Nong Ta was watching me.

'I've already been through everything before you came, Khun Tarrin,' she said. 'There's really nothing there to tidy up, just the bottles. You should take them. He would have wanted you to.'

I shook my head. 'No, I think they should be cremated with him. He enjoyed his Mekong and will probably need a drink when he gets wherever he's going.'

'Do you really think so?' she said, and burst into tears. I made no movement towards her, at once surprised by her sudden outburst and at a loss as to how I might comfort her. 'It's so sad, Khun Tarrin,' she said between sobs. 'Who'd have thought it, a heart attack at his age? He wasn't that old, was he? His poor wife... how is she going to cope all alone?'

She concentrated her gaze on the bottle in my hand and, regaining her composure, said: 'So many people have heart attacks these days. It's the stress of our modern life. Maybe with Achaan it was all the drinking.'

'Yes, maybe it was,' I said weakly as I put the bottle down.

I proffered a box of tissues, which she took and, dabbing at her tears so as not to smudge her careful makeup, asked: 'What will you do now?'

I had despised Achaan while he was alive, and in his death I despised him even more. Khun Lung Serm's request for me to write an obituary seemed absurd. But I was trapped by duty. Having accepted the job in the first place, I had to see it through to the end if the circle was to be closed. I took another long, disgusting pull at the bottle of Mekong. Then, crossing to the window, I looked out across the jagged Bangkok skyline. The clouds overhead were already merging with the rising smog of the streets into a dirty grey. The high-rises looked especially ugly in their disproportion to the flat landscape beyond. What had I achieved by coming back?

I suddenly felt a strong need to speak to Pi Taew. I needed to tell her how sincere my feelings were, and I needed her to reassure me that what we had said to each other during those afternoons in Chinatown was real, not fake like the kind of truths spouted by people like Achaan Vira. I wanted to tell her that everything was going to be all right if we could just dare to be guided by love and integrity. My heart was pounding as I keyed in the numbers of her mobile. But her phone was switched off and I rummaged in my wallet for her home number, obtained from the mutual friend who had introduced us with the promise of an introduction to Achaan. I hesitated for a moment knowing that she might be taking her children to school at that hour. But I was desperate and I didn't care if her husband answered the phone.

'Khun Taew mai yoo. Khun Taew's not home,' said the female voice at the end of the line, her thick southern accent heavy with suspicion.

'Has she taken the children to school?' I asked.

There was silence.

'Has she gone to work?'

'No. She's gone away,' the woman said after a long pause.

'Gone away? Where?'

'To Rayong. Excuse me, but who are you?'

'I'm sorry. I'm a colleague at the company. It's about work. It is important I contact her.'

Another silence.

'Maybe her husband might know?'

'Her husband? What husband? What are you talking about?'

'Khun Taew told me that her husband was about to be promoted.' I don't know why I said this; it was an unnecessary detail.

Again I was met by the humming silence.

'He is a policeman, isn't he?'

'Policeman? ... Who are you?'

I ended the call, lightheaded, and then I turned and saw that Nong Ta was looking at me. I had begun to laugh, softly at first, then hysterically and uncontrollably.

I wrote two pieces for Khun Lung Serm.

The first was this:

> There is only one way to honour the man who claimed to tell the truth, and that is by telling the truth. Achaan Vira's life came to an abrupt end late last night when he was stabbed to death by a young man who made his way to his house and into the bedroom where he and his wife were sleeping. This man, Achaan's lover, was high on drugs. He told Achaan that he had come to punish him for being disloyal, or for being cruel, or simply for being a liar. This is

my guess. Perhaps it was something worse. He may have told Achaan's wife that her husband had never had any pleasure with her, that he had always loved men, that his life with her had been a lie, that in private when the doors to the world were closed he was a man who had always only liked men and that his marriage to her was a sham. I don't know. In any case this young man stabbed Achaan three times in the chest with the knife he was carrying and slit his throat before ending his own life. Achaan's wife called her brother, a senior policeman who made sure a team went round to clean up the mess and the stains and that the whole episode would never be reported in the press. Those with power in our country can do that kind of thing. We then swallow what we have been told. I did not know Achaan well. We worked together. I disliked him intensely and he probably felt the same way towards me from what I could tell. To me he stood for many of the things that I find disturbing about our country, perhaps most of all the hypocrisy. Our programme was called 'Pood Kwaam Jing', 'Telling the Truth'. It would be nice if someone would begin to do so in our country. Achaan Vira certainly did not. But I do not want to vilify a dead man and, besides, as a Buddhist he will have to deal with his karma in another incarnation. All I can say that is good about my acquaintance with him is that despite my rejection of all his ideas and all that he stood for I am, nevertheless grateful to him in a strange way because his death and the terrible manner of his dying has woken me up from a bad dream.'

The second piece was this:

It will be public knowledge by now that Achaan Vira passed away late last night as a result of a heart attack. As a result we will now have to stop broadcasting 'Pood Kwaam Jing', which, in its short time on air has become a household phrase. As director of programming for Radio 9 I had the pleasure to work closely with Achaan. He was both a man of principle and a colourful figure who had come through a long career during which he served the country in many capacities, as a politician and a servant of the people. His last job as a radio broadcaster was no less important and his contribution to constructive debate on national and social issues will be remembered and appreciated for a long time. Achaan Vira made this programme popular because of his frank interchange with his listeners, his sensitive advice, as well as his ironic comments on world affairs. We are sorry to have lost such an important personality and we extend our condolences to his wife, who must feel his loss very deeply.

I printed out both versions and placed them, side by side, on my desk. My work was done, my account closed. I had started with nothing and ended with nothing. I was again on my own. I would start again. I would stop hiding behind alcohol. I would stop kowtowing to those I did not respect. I would stop seeking love where there was none to be had. I would finally do what I really wanted to do, which was to become a writer. As these thoughts came to me I felt an absurd surge of confidence, even though I did not know if I had the courage and the determination to follow the path that now looked possible.

In the corridor I passed Khun Lung Serm. I told him that I had left the prepared eulogy on my desk, but I did not put my hands together and I did not say goodbye. I was done with meaningless, insincere etiquette. I couldn't do it any more. It was nothing personal. I could sense his eyes boring into my back as I walked away down the corridor. I thought to myself: *Fuck you! Fuck Achaan Vira! Fuck Pi Taew! I'm done with this reality*. In that moment I felt liberated and clean.

Chapter Three

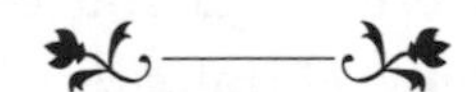

After rereading 'Achaan Speaks the Truth' Clare searched in her filing cabinet and found the handwritten notes she'd taken as she'd edited the piece, and which she'd used for her subsequent pitch for the story:

How much of this is fiction? Achaan Vira – real?

The hypocrisy and double standards is well captured – a real ogre. If this is autobiographical then what is T's agenda?

Tracks Thai history obliquely, and reflects the corruption without making a meal of it. He makes himself part of the problem – that's good.

The Pi Taew figure. Could she have hurt him that badly?

They were only casual lovers. How can he have projected so much on to her?

Intricate doodlings covered the rest of the page; scratchings, odd figures, abstract shapes, designs reminiscent of a Paul Klee painting. Seeing these marks again surprised her. They were the ideograms that emerged whenever she was provoked into a process of thought that could not be adequately expressed by language. Even now she could not decode them. But she recalled they were about Bangkok. Then at the bottom of the page she had scrawled: *The man's illness* and underlined it three times.

Returning to these notes she remembered exactly why Tarrin's autobiograhical story had stirred up her connection to the city she had wanted to forget.

Later that same evening, between sips of gin and tonic, Clare wrote an email to Tarrin reminding him that it was she who'd edited 'Achaan Speaks the Truth' for the anthology, and who was now inviting him to submit an essay on the tsunami. 'We would be honoured if you would consider...'

After three days she received a reply and an attachment: 'The Coming Suffering', an essay explaining the unpreparedness and the negative ecological factors adding to the scale of the tragedy. It was a straightforward analysis of what had happened in the tsunami backed by solid data from various experts. At the end he warned that Bangkok was also at risk from an impending disaster, either a flood or an epidemic, or both, and that the authorities were again doing nothing to prepare for it.

The essay was duly published in a collection called 'The Wave of Destruction'. Clare sent a thank-you note along with his author copies. Tarrin replied, saying that any royalties should be sent to a Thailand-based charity that was chanelling money to the victims of the tsunami.

Almost as an afterthought he added that he had written a novel about Bangkok being flooded, set in the very near future, the theme he had alluded to in his essay. She immediately wrote to him saying she would be glad to read it and would he send it as soon as he could, which he did.

She could not recall exactly why it took them so long to make a decision about it. Harriet, their assistant and a recent Cambridge graduate, was the first to look at the text, which was then passed up to Clare with favourable remarks. Clare's own notes merely said: *Very promising but needs work*. After that it was Paul's turn. At some point before Paul had got round to finishing it Clare received a short message from Tarrin saying he had just signed a contract with Deva Books in Singapore who were going to publish the novel.

She had wished him luck, never thinking their literary paths would cross again.

Chapter Four

London, May 2011.

Clare has a moment of déjà vu.

She, Paul and Mira are at the table in the small conference room at Spitalfileds. The horse chestnut outside the window is coming into bloom. She feels they have been there before, sitting in the same position, but she cannot quite be sure. It is unsettling.

Paul is saying: 'My, how time flies.' He is referring specifically to their recent annual trip to the Geneva book fair. 'From one year to the next. Where has the in-between gone?'

'It's called the Bardo in Tibetan. It means Death.' Mira offers this without looking up from her iPhone or changing her expression.

'Ah, yes, I know,' says Paul tetchily.

He has been telling Clare that at the book fair they had met Sophie Liu, the owner of Deva Books.

'They're folding, like so many. Even in Asia. We'll be next if we're not careful,' he says pensively. 'Anyhow, she offered her whole stable to us. I told her we couldn't afford it. But there are a couple of authors I've been wanting to pinch off her for years.'

Tarrin Wandee's name is mentioned along with two other Asian writers.

'Sophie said your old friend's got a novella in the pipeline. That's what she called it, anyway. They've already edited it. She was wondering if we wanted to take it on. His last novel did rather well, apparently. She said they've also found out that he's seriously ill – cancer, something like that.'

Clare had asked Mira to go to the book fair in her place after mumbling the excuses that over the years Paul had come to accept with an upward roll of the eyes. But she knew he was grateful each time for the justification of spending a few days with Mira without the furtive guilt-ridden angst that had marked their affair from the beginning. The European book fairs had become their protracted *cinq-à-sept*.

Mira had now been elevated to the rank of assistant editor and Clare knew that the day was approaching when Mira would take over her own position, unless, for some reason, Paul's feelings for the girl turned sour. After Geneva she'd noticed that there was a tension between them that had not been there before, and she sensed he was going to confide in her before too long about the agony he was going through by being unfaithful to his wife.

Clare was glad that she still felt interested in what was going on between them. In the not too distant future she would not even know who they were, nor care what they did to each other.

When Paul and Mira were away, Clare had two appointments at King's College Hospital. In the first of these she had submitted herself to the torture of the MRI chamber, and on the second she'd been given the results. They had revealed the tell-tale signs she had feared, and which the consultant had expected.

'It's very early days,' he said as he showed her the image of her brain. 'Always practically impossible to tell anything about the timing.'

He talked of the options, the medicines that were now available to slow down the process of atrophy, the mental exercises that might help. They were going to monitor her regularly. He told her there was no hard-and-fast cure just yet. They were waiting for a breakthrough, but she shouldn't pin any hopes on it. She was grateful for his frankness.

'But the most important thing is to prepare the people around you. You're going to need their support.'

It was his final piece of advice before her next appointment.

That evening, drink in hand, she sat in front of the TV not paying attention to the screen and digested the consultant's words. The gin did little to calm her. Fear produced waves of adrenaline making her more alert and focused than she'd ever been.

She could not get the word atrophy out of her head, and she realised that up till then she had used it as a poetic term, a metaphor. But now she was faced with the physical condition it described and the full horror of what lay ahead.

'Prepare the people around you,' he'd said.

She did not tell him that she had no one close left except Joanna. Annie, with whom she had shared five years of her life, had moved back to New York when they'd split up ten years earlier. They had not parted on good terms, or kept in touch, and she did not want to try now to renew something that belonged to the past. Besides, Annie would never have coped. She was a princess who liked to be looked after, and waited on. Clare then thought of her other lovers. But there was none among them she wanted to see again. The consultant's advice made her realise, yet again, that she had not invested in any lasting friendships. She had no support group to go to. She had led a more or less a solitary life, marked by the steady routine of work and nurtured by culture and literature rather than the everyday drama of reality.

In the end there was only Joanna. But Clare decided she did not want to prepare her sister just yet, and to become a burden to her too soon. So she would prepare no one, and not even mention her illness until she felt she was no longer able to look after herself. She still had a margin of time.

'I must make every day count,' her father had said when at the same stage. 'I must tie up the loose ends.'

For him, a practical man, this meant making sure his finances were in order and the will he was leaving behind was clear and fair. But in her case there was nothing external to sort out. She'd already seen to it that everything would go to Joanna and her nephews; the flat in Belsize Park, her small collection of rare first editions, the bits of jewellery inherited from her grandmother.

There were no loose ends to tie up, only a knot deep in her heart that needed to be unravelled before she could let go.

Chapter Five

It was not until the end of August, just before they all went off for their two weeks in the sun, that Tarrin replied to Paul. It had taken time for him and Deva Books to sort out the accounting issues and part company. Tarrin's novella was attached to the email. It was called 'Curtain of Rain'. Paul had his secretary print it out for Clare, knowing how she preferred to work from hard copy rather than a computer screen. The next day he left for Cyprus with his wife and three children, Mira for India to see her relatives, and Clare for Godalming, where, between long, pleasant walks in the woods, she read through the script of the author she'd never met, or Skyped with, but who she felt she knew. What she found in his stories both surprised and disturbed her. Yet again she found herself plunged back into the past.

Nothing neurologically abnormal happened to her during those sun-filled weeks in the countryside except that she forgot her nephews' names a couple of times, and covered up by calling them 'darlings'.

In her appointment with the consultant in early September she told him how well rested she felt and that there had been no significant signs of deterioration in her memory. In fact, nobody in her family had noticed any change in her.

He commented sympathetically: 'The medicine is probably doing its work for the moment. That's good news. And have you asked anyone to help you?'

'No, not yet,' she replied, and he raised his eyebrows.

'It's not going to go away you know, Clare.' He'd said this very gently and she nodded. He had not meant to undermine the confidence she had been feeling. Nevertheless, after the consultation she found herself trembling once more.

The Monday they were all back in the office Clare went immediately to see Paul. After exchanging accounts of their holidays she told him:

'I want to go to Thailand.'

Paul looked startled and hesitated for a few moments.

'But you hate flying.'

'I know. But I have to go and meet Tarrin.'

Paul still looked puzzled.

'I have to ask him something,' she knew this sounded both vague and unconvincing. Paul shifted in his seat.

'Do you really need to go all the way to do that? I mean...'

There was no way Clare was going to explain.

'I'm owed another week's leave. I'll pay for myself if the company doesn't.'

In the end he arranged a business-class ticket for her and a good hotel by the river. Clare guessed that he had found a way of persuading the directors that the trip was necessary. He owed her, yet at the same time she understood that she was almost beyond her shelf date. This was going to be part of her golden handshake.

Part Two

by Tarrin Wandee

I
Mae Jom and the American

Bangkok under a white haze: in all directions, a jagged skyline of irregular glass peaks that pierce the bleached canopy; unfinished buildings exposing naked concrete, rusting rebar, cranes perching on top like idle crows waiting for the next dubious downpour of construction cash.

Mae Jom feels threatened by the chaotic unfamiliarity of the city. She has been gone forty years; it could have been four hundred. She recognizes neither place nor pace nor proportion. Gone is the wide, sleepy sense of space. She searches for a reference point, but cannot establish her bearings. She looks again at the neatly written directions on the damp scrap of paper in her hand, but for a moment they seem incomprehensible. Mae Jom tightens her grip on her small suitcase and walks uncertainly in the direction of the signposted SKYTRAIN arrow, falling in behind a confident-looking couple who seem to know how to find the platform.

She stays close to the pair – the man a farang in his early thirties, about six feet tall, lean and broad shouldered and wearing a black T-shirt decorated with an image of a huge green ganja leaf, blue knee-length denim shorts roughly cut from an old pair of jeans; the girl by his side, even in her extra-high heels, is barely half his height and deceptively young with light makeup and smooth, taut skin. She is wearing a classic hostess outfit more suited to the night. Her tight, low-cut, short and fitted red dress pushes her small breasts upward and together, emphasizing cleavage. She has a pink plastic handbag tucked under her left arm. She is dark, with shiny black hair so long it reaches the small of her back.

Together, these two create a spatial island that makes Mae Jom feel safe as she hovers at its edge. People pretend with half-smiles not to notice the farang on the increasingly crowded platform, or the girl, whose face, devoid of any expression, brings back strong and unwelcome memories for Mae Jom. It is impossible to guess what the girl might be thinking or feeling. The farang is standing close to her, the back of his right hand touching her bare upper arm below the shoulder, stroking upward with the most delicate and languid of movements. There is a look of melting tenderness and adoration in his eyes, a contentment that excludes them from the rest of the world. The girl does not respond but Mae Jom now recalls such intimacy and her own arm begins to tingle, touched by a ghost.

Had it not been for the incongruous couple, she might have felt self-conscious among the smart city people milling on the platform because she is aware that she looks different from them. It isn't just because of what she is wearing – for the journey to the capital she chose the clothes she reserved for the temple, weddings and funerals: dark brown trousers, a light cotton blouse, tired slip-ons. It is her grey hair, cut short in the traditional way, her air of awkwardness, her cheap handbag, her skin grilled dark by the fierce sun of the northeast countryside. Mae Jom feels grateful for the invisibility accorded her by the presence of the couple.

A gust of warm air precedes the arrival of the sleek train and Mae Jom is swept up in the crowd as they board. She must concentrate now on the task at hand, to get to the shop-house where she will be staying with Nai Pot, her young cousin, who has made all the arrangements. His place is far from the bus station, on the other side of the river at Thonburi, and a long walk from the last stop of the train. He will find her at dusk when he finishes work.

'Be careful,' he warned in his instructions. 'Bangkok now is nothing like the city you knew. It is no longer so friendly and easy-going. There are many bad people waiting to take advantage of newcomers from the countryside or from the borderlands.'

Mae Jom is puzzled by his caution. She senses no threat, nor can she read danger in the blank faces of her carriage companions, but she concedes that what she sees through the windows of the fast, near-silent train intimidates her. At Siam Square, where she must change for a connecting train, she cannot reconcile her memories with the vast glass-fronted shopping centre on one side of the platform and the huddle of crumbling buildings on the other.

Despite the proliferation of high rises, and the bridge itself of only recent construction, the river is familiar. The sight of the greenish brown water with the sunlight dancing on its surface reassures her. Looking down she smiles as a river taxi, its deck crowded with tourists, crawls against the current towards the old quarter and the Temple of Dawn. She watches a long black barge weighted down with sand and gravel and towed by a small tugboat pass under the bridge, jouncing in its wake an old wooden ferry Mae Jom remembers taking from one side of the river to the other, standing in its stern hand-in-hand with someone she scarcely knew.

Nai Pot's long walk turns out to be just twenty minutes, which amuses Mae Jom – her daily walk to the fields takes longer – and she thinks that city living has made her cousin soft. On this side of the river she feels more at ease, the shop-houses and tree-lined streets look more like the old Bangkok of her memory, but her nose wrinkles at the canals, once clean and filled with bobbing boats and sleek naked children diving from the steps, now black and bubbling with debris.

The smell of Bangkok has bothered her since she left the bus, but it is not until she alights at Wong Wien Yai, the station on the big roundabout of Thonburi, that all Mae Jom's senses are open to what the city has become since she was last there: so many people, as if half the country had moved to the capital; the traffic, when once the streets were so empty that taxis could race each from around the city, inching deafeningly forward in a choking miasma; the pavements clogged with stalls spilling over with clothes and shoes and watches and gifts, all at a special price for those who stop to look or buy and making it that much harder for Mae Jom to pass.

She makes her way slowly and patiently. She is in no rush – Nai Pot , a driver, has still a couple of hours work ahead of him – and her senses are both confused and excited by the vibrancy. But after a while the car fumes and smoke from big woks and the stench of stale piss and rotting garbage are making her nauseous. She wants to escape this cacophonous thoroughfare.

Mae Jom stops for a moment, centering herself; Nai Pot's instructions are clear. She must go directly to the market, no detours, and so she continues at her snail's pace, the white heat of the afternoon shimmering around her. She cannot help but notice the proliferation of mobile phones pressed against every other ear, young and old. She sees more than a few people talking to themselves as if quite insane, gesticulating with both hands, oblivious to the world around them, until she spies the earpiece. Even as they sit at the food stalls eating their noodles they continue to talk, taking no notice of anyone else. There were several mobile phones in her village, but those who used them did so quietly and politely, conscious of the people queuing to make their own calls. Perhaps it was the Bangkok way of blocking out the noise, which manifests itself to her as a wall of sound; voices amplified, shrill whistles from parking attendants, music pulsing from all directions, blaring car horns, and the ineffectual wailing of sirens from ambulances and police cars anchored in traffic.

Ahead is the marketplace, and Mae Jom turns down the cool, dark, narrow lane running off it, gasping with relief, soaked as if she has emerged from a whirlpool. Nai Pot's shop-house is close and she finds a stall selling iced coffee. She buys one and sits there, sipping and waiting for the sun to fall and the weather to cool down.

That evening Mae Jom lies in bed listening to the dull, thumping rhythms of the loog toong band playing in the market and the doleful voice of the singer over the microphone lamenting lost love. She feels exhausted from the journey down and the trek across the city. Now in the trancelike state produced by the music she keeps returning to the farang and the girl in the red dress and she recalls the love light of affection in the tall man's eyes and suddenly, with searing, intense pain, a memory shoots through her consciousness.

She sees her eighteen-year-old self, standing on the crowded platform of Hua Lampong station. Her hair is long and soft, held back in a loose ponytail. She is wearing a white cotton sleeveless top over a pair of blue jeans, waiting for the train to Udon. A small, tan-canvas suitcase is by her feet. A young farang with military short hair and wearing a Hawaiian shirt is holding her hand, lightly, in his. Her palm is damp. She does not want to leave him. She sees the way he is looking at her. He is telling her how much he loves her and she feels frightened, unsure if she can return the intensity of his love, so she turns her head and looks down the track that winds its way out of the city. A voice crackles from rusting speakers and she hears the distant whistle of the train. She knows she has run out of time.

Mae Jom does not want to be in Bangkok but she cannot argue with the circumstances that make the journey necessary, nor can she neglect her duty, her obligation. When she was younger, she had obeyed without protest. Now, in her fifty-eighth year, Mae Jom had permitted herself a moment of resistance and displeasure as she set off from Udon, refusing to return Nai Gawb's wave. She watched him turn away with a look of resignation and walk head bowed into the early morning mist. As the bus pulled out, she glimpsed in the window a young woman with long black hair and innocent eyes and thought it was a hallucination, but it was only the girl in the seat beside her, half standing, waving at her boyfriend. Then she felt sorry that she had been so mean to her husband. Nai Gawb had not meant any harm. And life was too fragile for resentment. It was possible that she might never see him again. She always believed that everything was held together by links that could be broken at any moment.

It had been nearly four decades since she'd last journeyed away from the village. In the interim the only public transport she had taken was the country bus, irregular and inconvenient, or shared country taxis: old estate cars rescued from the Bangkok scrap heap and rebuilt to transport ten or more passengers and all the stuff they carried with them. Yet here she was aboard a brightly painted double-decker that seemed to fly along the highway. Nai Gawb had wanted her to take the bus, rather than the cheaper train, to Bangkok. The non-stop service boasted reclining seats, onboard toilets, and videos that were shown on the screen all night long.

During the anti-government protests in Bangkok, he had watched as whole convoys sped out of the countryside towards the capital, paid for, it was said, by local businessmen who supported the ousted Prime Minister Thaksin Shiniwatra, demonized in the central provinces but still a hero in the Isaan region. A 'champion of the poor', they had been told to call him, though Mae Jom had seen no evidence of this. Nai Gawb had wanted to join them, but she thought he only wanted to ride the bus and see Bangkok, all expenses paid. He had been given his red shirt and his bandana and was ready to do whatever they'd asked, but his mother, Yai Da, was sick and begged him not to leave her. So he swallowed his disappointment and stayed behind, as he did again when Mae Jom was summoned.

'Please take the bus,' Nai Gawb had insisted. 'We can afford it. Then you can tell me what it was like. And I want to know everything about the city. I want you to keep your eyes open so you can come back and tell me what's been happening there since we left.'

Mae Jom did not want to argue with her husband, nor did she have the strength to fight Yai Da, for whom, with grim irony, she was making this trip. When the doctor at the hospital in Udon had determined there was nothing much more that medicine could do, he had sent the old woman home. She accepted without complaint the fact that she was soon to move on from this world. Her only wish was to see her granddaughter for the last time so she could die in peace. How could Mae Jom refuse? Yai Da had looked after the baby as if she were one of her own, as she had done with Mae Jom, whose own mother had died and whose father had vanished before she'd had time to remember him.

Yai Da's husband had died young and left her a smallholding of paddy fields and a mulberry grove, but the soil was not so fertile and demanded hard labour to sustain the family and carry them through the increasingly frequent drought years. It was a hopeless life, where the only winners were the middlemen who dealt with the market and the entrepreneurs in the capital who fixed the prices. The dream of every young person was to escape the fields and find a better life. Those who did manage to leave went to Bangkok. There they would find work and send money home, and save some too so that one day they would come back rich, or relatively so. In Yai Da's family, a cousin, Pi Lien, had several years earlier taken her small boy, Pot, and found work as a cook in a wealthy household. Her remittances home were sometimes all there'd been to sustain the family.

Nai Gawb had gone to Bangkok when he turned twenty. He had seen other villagers coming back with money to set up shops and businesses and buy farm machinery and fertilizer and pesticides and pay off the family debts. It was 1971 and everyone believed the

Americans would fight in Vietnam forever because their politicians were too weak to admit failure, though fewer Americans were now doing the dying – 'Vietnamization', it was called. Though there were fewer dollars to go around, people said Bangkok was still thriving and many like Nai Gawb were ready to believe that escape was still possible. Otherwise, in the white light that hung over the rice fields, dreams would remain just that.

Nai Gawb's early communications were positive and full of amazement at what he was discovering in the city. There was no problem finding work, he'd said, and what he had found was even better than what he had imagined. According to him there was an unbelievable choice: waiting at tables in restaurants, driving taxis, labouring on construction sites, the entertainment industry, anything as long as you were willing to work hard.

He told them that he had struck lucky, and that through a contact he had made he would soon be a taxi driver. Yai Da, knowing how her son liked to fantasize, did not believe him at first. But she put aside her worries when the first banknotes arrived. She told Mae Jom she was proud of her son. After hardly a year, Nai Gawb wanted Yai Da to send Mae Jom to Bangkok. There was room for her where he was staying and work to be had, he explained, and together they could save more. And, besides, he needed someone to look after him because he found it hard living alone in the capital: he worked long hours and often through the whole night; he had no time to wash his clothes or clean the place where he lived; he had no one to do the shopping for him. He needed her help. Yai Da said she would think about it. She was nervous about being left alone, but she knew she had to agree to her son's request and told Mai Jom: 'Go ... look after Nai Gawb. He's hopeless on his own. Make sure he's safe, that he's eating right. Find yourself a job and send money back if you have enough to spare. But come back to visit when you can. Don't let the city swallow you up like the Naga.'

Mae Jom did not want to go. Nothing she had heard about Bangkok excited her, despite her understanding of their lowly situation as farmers. Life in Isaan was hard, but she was content to walk the fields and watch the almost imperceptible passage of the seasons, to feel the mist of February mornings and the warm monsoon rain, to prepare the silk for weaving and to help with the harvest. The stories people told about the big city filled her with a vague sense of anxiety, as though something bad would happen to her there. But she could not disobey Yai Da, nor could she deny Nai Gawb's request for help: they were her family.

Nai Gawb fetched her from Hua Lampong station in his pink taxi and they drove through the streets with the windows open, the wind blowing through their hair. With the fresh eyes of a country girl she saw everything with a bright clarity: the solid buildings; the wide black tarmac roads; the giant billboards; the pavements full of people in expensive clothes; the shiny cars. She felt intimidated, and acutely self-conscious, but Nai Gawb's cheerfulness infected her and soon they were talking and laughing as he explained how Bangkok functioned and shared with her his hopes and eager expectations.

Their destination was the Makkasan district, where many from the provinces started their urban odyssey. There he parked outside a wooden shack. 'It was the cheapest I could find,' he said, and led her into a dark, musty space and turned on the single electric lightbulb. 'When we start earning more money, we can move to a bigger place, maybe a room in an apartment building if we're lucky,' he added.

Mae Jom hoped he was serious. The space she entered had been divided with makeshift curtains, with a small bathroom and toilet to one side and a cooking area to another. The main part of the room was a mess of empty cardboard boxes, unwashed plates and cups, dirty clothes. Yai Da was right: her son couldn't look after himself. In the far corner Mae Jom noticed a newer piece of cloth strung up to screen a low bed made from discarded wooden crates, and a scavenged wooden cupboard for her clothes. She was grateful for the effort, but she couldn't help notice that the rest of the room had only the barest of furnishings and that its focus was a large shiny portable cassette player. Nai Gawb was proud of this machine and at once demonstrated its power by putting on a cassette of a loog toong band and turning the volume so high she thought the roof would cave in. He looked hurt when Mae Jom covered her ears shouted for him to turn it off.

On that first evening they ate at a noodle stand at

the end of the lane. Nai Gawb drank a large bottle of beer and talked non-stop as he slurped down his food. Mostly, he talked about how they were going to get rich. She listened quietly and respectfully, and tried to share his optimism. But she remained unconvinced. Still, there was no turning back. She had to try her best and put aside her misgivings.

On the lumpy mattress listening to Nai Gawb's snore from beyond the curtain and the hum of the city beyond the shack, she prayed for courage to overcome her sadness and apprehension. When it began to rain, the sound of the raindrops hitting the corrugated iron roof, at first gentle and then thrumming, she thought of how the dry fields back in the village would welcome it, and when bullfrogs began their echoing chorus she smiled because she had not thought that they, too, had come to the city to seek a better life. She slept until dawn, when she rose quietly and then went out to explore. To her pleasant surprise she found a small lake nearby that was full of fish. One of her new neighbours was casting his net into the rays of a bright red sun skimming the surface of the water.

It was an oasis.

Mae Jom soon found work as a waitress in a large two-storey restaurant in Siam Square, which would grow to become one of Bangkok's main shopping and entertainment areas, but at that time was a jumble of demolition and construction. Low-rises were sprouting up next to tracts of wasteland where mangy dogs lay sheltering from the sun under huge trees. Mae Jom had never been a waitress before; indeed, she had never even eaten at a table or sat in a dining chair. She knew only the floor, so the other waiters and waitresses guided her through the first day. The pay was low, but the tips they shared equally at the end of each evening gave her more money than she had ever known.

Hers was a simple routine: she would get up early, do the cleaning and washing, prepare rice gruel for herself and Nai Gawb, and leave late in the morning. She would come home in the dark, tired but a little bit richer. Before turning in she and Nai Gawb would share what they had been up to and what they had witnessed in the course of the working day. Nai Gawb was happy to have her looking after him, and to have someone to tell his stories to, and even though he could see she was not totally at ease in Bangkok he was sure it was only a matter of time before she would leave the country behind and adapt to the new rhythms. On days off they would visit their cousin, Pi Lien, and her son, Pot, and together go to the park or to the river for a meal and to swap news from the village. Little by little Mae Jom began to share Nai Gawb's dream.

American soldiers on R&R were the lifeblood of Bangkok and the city took seriously their needs for recreation, if not rest. Near Makkasan on New Petchburi Road there was the area where mostly black soldiers liked to hang out. James Brown and Otis Redding blared from clubs and bars that never seemed to close. In the lanes behind there were shacks where the parties would go on through the night, with the smell of marijuana hanging heavy in the air. Further up on Sukhumvit Road the action was to be found around the Grace Hotel, where the white boys stayed and where the mamasans made it their business to make sure their clients were kept happy.

Mae Jom's restaurant was popular and catered to both locals and foreigners: Western food on the lower floor; Thai and Chinese cuisine on the second. Foreigners who ventured up the stairs tended to be those who lived in Bangkok, either for work or lifestyle: businessmen, journalists, hippies who had landed in Bangkok and who were caught in its dizzy spell, endlessly making plans to leave for other parts of Asia but never doing so.

She had been working for more than a year when a group of American soldiers on leave, already high, stumbled up the stairs one evening. Mae Jom had been serving a group of loud, heavy drinking regulars. Among them was a tall farang woman, with translucent white skin and long brown hair in a short skirt and frilly top. The woman's lone, feminine, presence in that raucous male company intrigued her. Mae Jom led the soldiers to the adjacent table. Within minutes, the two groups were arguing and, although she could not understand what the argument was about, she knew from the intensity of their expressions and the way they were throwing their words at each other that it would possibly end in violence. She caught the manager's eye but he pretended not to notice and shuffled off to the far side of the restaurant to exchange greetings with some local shopkeepers. Other customers shifted uneasily in their chairs, ready to run, which would have meant no tips for Mae Jom, unpaid checks and a mess of scattered food and broken furniture for her to clean up when the brawl ended with the whistles of military and local police.

At one point Mae Jom saw the soldier who'd been the most aggressive reach into the back of his trousers and draw out a handgun. 'Just like in the cowboy movies', she thought. She froze like all the other customers and resigned herself to the inevitable when the farang woman jumped up, put herself between the two red-faced groups of men and shouted at them all to 'sit down and shut up, the lot of you' and used some more English words that didn't need translation. There was a fierce look in her eyes and to Mae Jom's surprise all the men obeyed and for a moment, save for the noise of traffic outside, the entire floor was silent. It was the first time Mae Jom had seen a woman command men.

Order was restored, conversations resumed, plates clattered. Mae Jom took a deep breath and approached the soldiers, holding her order pad to her breast as if it were a talisman. They ordered pitchers of beer and chose the spiciest Thai food on the menu, Isaan dishes that even some of the locals found too strong.

All through the meal one of the men, whom Mae Jom had noticed hanging back from the fight, kept staring at her and, though embarrassed, she could not help but look every now and then to see if he were still watching.

He had short, light-brown hair and eyes green as fine jade. He was quieter than the others and had a sadness about him, even when he smiled. She felt her face grow hot as she blushed and lowered her head. He was still smiling at her when she looked up again and she busied herself clearing plates and dabbing at damp patches of spilled beer as his friends engaged in an amiable tussle over whose turn it was to pay the bill.

He followed her from the table and, with the bravado of someone who has nothing to lose, stopped her. With hand signs and words she did not understand but whose meaning was clear, he asked her to go out with him the next day. Without quite knowing why, she accepted.

She swapped her shift with another waitress and went with the soldier to Dusit Zoo in the morning and then the cinema, where, in the dark that smelled of cigarettes and cheap perfume he kissed her softly on the lips. They ate at a Mexican restaurant and for the first time in her life, Mae Jom was served at the table by a waiter. Afterwards they walked down the crowded road, without saying a word to each other while he gently held her around the waist, and then they went back to his hotel.

Mae Jom would try to explain to Yai Da that she did not feel any shame going with him. He offered neither money nor gifts. His advances were direct, and she could not understand why. It wasn't so much that she was attracted to him but that she was flattered that he, a white man, should take any interest at all in a plain dark-skinned Isaan girl. If he'd wanted only sex he could have had it at the wave of a hand, and with far prettier girls than she was. There was innocence about him, and she felt it would be wrong not to yield.

The next four days were spent with Johnny. He would come to the restaurant and wait for her to finish and they would go to a bar and have food and beer before heading to his small hotel. It seemed as if she were living in a bubble that floated over the city. She knew it would burst, but until it did she wanted to have fun and enjoy her time with him. Johnny took her to places she could never have afforded, showed sides of the urban life to her that she'd thought inaccessible.

She began neglecting the household chores and worried about what to tell Nai Gawb. She did not know how he would react when he found out she was spending her time with a farang soldier. When she did tell him, he seemed impressed that she, a plain-looking country girl lacking in skill and ambition, had landed herself a farang. When she told him what was going on he slapped his thigh in admiration.

'Play your cards right and you could be over in the States one day,' he said. 'Then you can send home decent money every month to Yai Da and we'll all be OK.'

Mae Jom knew of women who had married GIs and had gone to the States to make new lives for themselves. It seemed to be an ambition shared by many of the girls who came to Bangkok, though she had never thought herself among them. But once Nai Gawb had planted the idea in her mind it was hard to shake it off. His words awakened the possibility of another life.

She knew little of this young American soldier's background, except that he was from a place called Ohio. They continued to communicate through gesture and sound, which made them laugh and try even harder to reach each other. She had not spent time with any other man except Nai Gawb, who was more like a brother. She knew little about sex, her sole experience having been at fourteen with a boy little older than herself at a temple fair, a fumbling, clumsy few minutes driven more by curiosity than mutual attraction. Yet despite her lack of worldliness Mae Jom knew enough not let her heart be overtaken by any emotion that resembled love. She was ever mindful that perhaps all this man needed was the comfort of her body for a few days before returning to the war. And on the rainy morning that she saw him put on his uniform and wave goodbye as he climbed into the taxi, she could not tell if she would see him again. She was not sure if this made her sad.

Johnny did return after four months. He sought her out and insisted she take time off from the restaurant to spend the whole of his one-week leave with him. Mae Jom didn't try to explain that in Bangkok there were only two states of being: you were either working or looking for work and time off meant losing her job. But at the same time she was tired of the long hours on her feet and the rudeness of the manager. One of the girls had left the restaurant for a job in a new massage parlour in the Patpong district, a popular place for soldiers and tourists and where the pay was much better than what she could get waiting tables, even when the jar that held their tips was overflowing. All she had to do, the girl told Mae Jom, was to wash a stranger's body in the bath, dry him, powder him, prod him up and down the spine, stroke his legs and, when he became aroused, either use her hand to relieve him or, if he paid extra, find a pleasant place in her mind while he penetrated her body.

'You are no beauty,' her friend said matter-of-factly. 'Neither am I. And look, I have cleared all my family's debts, made merit at the temple and paid for my cousin to go to school.'

Though she was as pragmatic as her friend, Mae Jom was hesitant. The money was tempting; but there was a big difference between dealing with the bad behaviour of restaurant patrons to having sex with strangers. She also remembered Yai Da's words: 'Don't let the Naga swallow you.'

During the week of Johnny's leave they rented a room in a cheap hotel near Hua Lampong where the hippies stayed. He bought her clothes and shoes and paid for everything from a roll of crisp, clean dollars. He made sure there was always money in her bag and told her not to worry about finding another job because he would take care of her. Mae Jom let herself be drawn closer to him and began to see a more intense and troubled man, unfaltering though he was in his desire to be her carer and protector. There was a darkness inside him that struck her as dangerous and unhealthy.

They went on a trip up the river to Ayudhya with a group of other soldiers and their Thai girlfriends and at a shack by the canal on Thonburi they smoked ganja from a bamboo water pipe and watched the fireflies and the small canoes paddling by in the twilight. Later in their room she could not stop laughing. They strolled around the markets in the daytime and lay under the trees by the lake in Lumpini Park, and watched Hollywood movies and danced in small, dimly lit bars off Sukhumvit where the professional girls with their careful makeup and perfect lipstick sat on barstools in low-cut dresses waiting for customers to buy them cocktails and pay the management to take them to a short-time hotel.

Johnny talked a lot about the war; it was never clear to Mae Jom what he did as a soldier, only that it was bad. When he thought about it a strange cloud would cross his face. Often during the night he would wake trembling or yell out and she would hold him close and try to soothe him. She understood that he had come to despise those who gave the orders, that he hated the war, that he was afraid, that he missed America. He showed her photos of his friends and family and said she would meet them one day. Mae Jom memorized their faces and tried to read in their expressions what their lives were like. They were all so white. In her mind, she played at conversation; copying Johnny's clipped English, preparing for the day they would all meet.

On the fourth morning, Mae Jom woke and saw Johnny sitting at the end of the bed watching her while he smoked a cigarette. He said he loved her and asked if she would marry him. He looked so serious she could not help but laugh. He repeated it over and over again, as though to make sure she understood. She wanted to tell him it was all going too fast, that she wanted to think about it, that she had feelings for him she had felt for no one else and needed to understand them. But she didn't have enough English to express all these things. So she nodded and squeezed his hand tightly and saw the look of gratitude in his eyes. That day they wandered lazily through the narrow streets and laneways of old Bangkok. They visited the temples, rode a long-tail boat across the Chao Phaya and into the small canals where naked children jumped from the bridges into the green brown water, laughing and splashing.

Standing at the doorway and clutching a towel around her thin body, Mae Jom felt a sense of panic. Nai Gawb looked tired and agitated and said Yai Da was very ill, Dengue fever, and one of them had to go back to Udon. He couldn't go. He had to work. She had told Nai Gawb where she was staying with Johnny in case of an emergency, and so he had come. She knew she could not refuse Nai Gawb but she did not want to abandon Johnny, who had to return to Saigon the next day. She used all the words and expressions in her vocabulary to explain the situation to him. Johnny said he understood, that family came first, that he was sure everything would be OK. He took her to the train station and told her he would be back as soon as he could, but he did not say when. He told her again that he loved her and that she should wait for him.

When she returned to the village Mae Jom discovered that Yai Da was very ill and she stayed close to the old woman, nursing her back to strength. Each morning, she walked to the letterbox on the main road in expectation of a note from Johnny – she had given him the address – but none came and she resigned herself to his silence. She went to the temple and lit incense and prayed for his return. When she missed her period, she reproached herself for not being more careful. In the next village was a wise woman, a herbalist who took care of these things, but Mae Jom's instincts told her Johnny would be pleased, so she did nothing. There was still time, and he would be back soon. She would let him decide.

Back in Bangkok three weeks later Mae Jom found a new job in a piano bar off Silom, where she served drinks to young Thai businessmen. It was better paid than her job at the restaurant, but the hours were long, the work more demanding. She told Nai Gawb her news and he congratulated her, saying she should celebrate instead of worry. He was sure from what she had told him that Johnny would be happy and that she would soon be in America with him.

She thanked him for his encouraging words, but knew that only when she and Johnny were face to face would she know how it would turn out. Until then, all she could do was wait and hope that the next soldier through the door of the bar would be him.

It wasn't Johnny she saw. It was one of the other soldiers who had been with them on their trip up the river to Ayudhya. She couldn't remember his name. The frowning manager, who thought Americans to be crude and rude and, regardless of their money, did not encourage their custom, told her to be quick. The young soldier stood in front of her with a funny expression on his face, scratching his forehead as if he'd been stung by an insect and moving from one foot to the other before finally mumbling some words, repeating himself a little louder each time until she understood: 'Johnny's not coming back. Johnny's not coming back.' And then tears flowed down his flushed, boyish face.

There followed a period of silent desperation for Mae Jom. Outwardly, she was the same: quiet and efficient, working dutifully at the bar and joking with the regulars, doing the chores for Nai Gawb at home in Makkasan. Her grief could not be expressed with tears, but took the form of an icy, internal anger and deep determination not to go under, not to be defeated, and out of that came her decision to keep the baby in defiance of the death of Johnny, whom she had not had time to know, and who was now gone for good.

The bar owner was kind and sympathetic but as the months went by and her belly grew he became uneasy and told her he had no choice but to replace her. Though Nai Gawb seemed to be working harder and harder – she rarely saw him now – and making more and more money, he was becoming unreliable. She worried that he might not be there if she needed him. Many of the girls from Isaan, even the ones with husbands and partners, went back to their villages for the final few weeks and Mae Jom had money saved so she could afford to put herself in Yai Da's hands. Few mothers of working age raised their own children. It was a fact of being poor and she was resigned to it.

Mae Jom and Yai Da made their preparations, and the services of a respected midwife from one of the neighbouring villages was secured.

'You are small but strong,' she pronounced, 'Not too difficult, I think.'

Then, unexpectedly, news arrived that Nai Gawb was in jail. He had been caught dealing heroin in a police sweep. It cost Mae Jom all her savings to pay off the police. Nai Gawb was released after only a few weeks, by which time he was a shaking wreck. Mae Jom had been too busy to notice he had become an addict, too wrapped up in her own problems to realise how much he had changed, and she blamed herself for failing to look after him, for failing Yai Da, for letting the Naga swallow him up. They persuaded Nai Gawb to return to the village and to go to a temple near Nongkai where the monks had experience with addicts. There he would spend three months, with herbs that made him vomit and saunas to cleanse his body, with meditation and prayer to cleanse his soul.

When he returned home he was a different person, chastened and humbled; the easy-going optimist was dead.

It seemed logical that they marry; Tookta, a beautiful baby girl, would need a father and Nai Gawb would need a purpose. Yai Da and one of their cousins, who was a monk, thought it was destined. They did not return to Bangkok. It was 1973. The Vietnam War had ended, at least for the Americans, although Saigon was still to fall. Many people were making their way back to the countryside as the 'recreation' industry pulled down its shutters. Some of the men went to Saudi Arabia or Libya to do construction work and others were making plans to follow. Nai Gawb wanted to join them, but his fight with heroin had taken all his strength and spirit and he surrendered himself to the land and its endless cycle of planting and harvesting, drought and flood. Mae Jom was spent, too. As a couple they settled down to a way of life they thought they had escaped.

Memory fades with time, like the once bright colours of their peasant clothes under the bleaching Isaan sun. If they ever reminisced about their youth, it was with others who had been to Bangkok; all their tales were sweet and sour. Mae Jom never shared her sorrow over losing Johnny with anyone. Her brief time with him had been too private, too precious. The best she could hope for was to learn to live with the pain. Sometimes when she was in the fields or at the loom she would imagine what might have been if she had been granted a life with him. Would she have liked Ohio? Would Johnny have taught her about love? Would they be happy? On the last question, she was not sure. Mae Jom saw how the Isaan girls who had gone to America with their soldier husbands had changed when they made rare visits back home laden with presents for everyone. Outwardly they looked different with their western clothes and mannerisms, and when they spoke, they betrayed their absence by sounding slightly out of sync with the local language, which had moved on without them. Mae Jom detected in some a sense of superiority, in others alienation, no longer sure to which culture they belonged, in all of them an unspoken yet palpable sadness. Their prayers had been answered, but no amount of incense could sweeten what they received. Mae Jom knew that if Johnny had not been killed she could have been one of them.

Tookta was different from the other children in the village. Her skin was fair, her eyes green. She looked like a doll, which is why Yai Da had chosen her name. Mae Jom had a photo that Johnny had given to her the first time they'd met. He was wearing combat fatigues and held a beer in his hand. She kept it in a cheap frame in a drawer under her best clothes. Tookta took after her father; there was hardly anything of Mae Jom in the child's features, except for her thick and lustrous black hair. Nai Gawb was proud to be the surrogate parent of such a pretty child. Yai Da, who believed in the way bonds developed naturally between one person and another, would swaddle the baby in a cloth, carry her and sing to her until she went to sleep in the wooden cradle, which Yai Da pushed gently and watched constantly.

In the end it was Mae Jom who had problems with Tookta's appearance in the world. The months leading up to the birth had been stressful, and she had invested her energy in her own survival in the city and in her fierce determination to keep the baby, and in not caving in to her feelings of grief. After the birth, she felt a kind of inner collapse. Instead of the joy and satisfaction and fulfilment she had expected, a deep melancholy seeped into every corner of her being and, as she held Tookta and gave her milk and washed her little white body, she found that she was filled not with love but with pain. Tookta reminded her of Johnny.

Tookta seemed to sense this, and preferred Yai Da's comforting embrace to her mother's. Yai Da did nothing to discourage Tookta's affections, though she demanded that Mae Jom be accorded proper respect. As her granddaughter grew, she showed her all the country ways: how to cook and prepare food for those working in the fields, how to spin and weave silk from the cocoons gathered from the mulberry grove, how to plant and harvest, how to give alms to the monks, how to pray in the temple. Tookta responded by becoming a good-natured girl who was considerate to others and respected her elders. Everyone liked her. If she were at all aware that her loog kreung features distinguished her from the rest of the children, and that the villagers treated her as a kind of mascot, she did not let it affect her behaviour. She was a village girl, like the others, with the same hopes and dreams, though a strong streak of independence marked her out. She was usually the ringleader in making mischief.

When she finished her secondary schooling, marked out as a bright student, there was talk of sending her to the teacher-training college in Nongkai. But Tookta had other ideas and soon after her eighteenth birthday she left a note saying she had gone to Bangkok with a friend.

Yai Da was greatly distressed by Tookta's sudden disappearance; that her sweet granddaughter had secrets had never occurred to her. It did not take long for Nai Gawb and Mae Jom to discover the 'friend' was a boy from the next village who was known to the police. They thought seriously of going to look for her but did not know where to start once they got there. The only family they had in Bangkok was Mae Lien and her son, Pot. They would not be of much help in such a vast place.

They had to accept that they had lost her. Yai Da became angry with worry and swore she would give Tookta the beating of her life when she returned home. The three of them went to the temple and consulted a monk said to have the gift of foresight, but he could give them only words of reassurance. They paid a healer from a nearby village to send special energies to protect her. And during the first year she was gone they lit a stick of incense each day and prayed for her safe return.

Years passed without a word. Then Tookta came back. One evening she climbed off from the back of a country taxi with two huge suitcases, walked through the village in the fading light and into the family house, where she stood in the doorway and, to a stunned Mae Jom, Nai Gawb and Yai Da at dinner, hands frozen between table and mouth, bowed a wai and smiled as though she had been gone but a day. She gave no explanation or apology, and they asked for none. Yai Da was overjoyed and, instead of the promised beating hugged her grandchild to her, crying and laughing at the same time.

Tookta handed out presents like a politician before an election: expensive tops and lipsticks and soaps for Mae Jom; a woollen blanket for Yai Da; a colourful windcheater for Nai Gawb; an electric rice cooker for the house. She told them she had found work as a secretary in a company that paid her well. She was being allowed a short holiday. None of them pressed for details. They did not really want to know the truth. It was enough to have her there with them once more. And had there been any urge to find out more it disappeared when Tookta handed Nai Gawb a thick envelope full of cash and finally asked for their forgiveness.

She stayed for a week, helped with chores and generally behaved as if she had hardly been away. When she boarded the bus back to Bangkok she promised to stay in touch, which she did, calling from time to time, often with the roar of traffic loud in the background.

Over the next twelve years, Tookta visited three more times, always without warning, always bearing gifts and an envelope stuffed with cash, and always reassuring about how well she was doing in the company and in her job. On her last visit, Mae Jom thought her daughter to be unwell, in both body and mind. There was a cough she tried to hide behind her hand, and blemishes on her skin. Mae Jom had seen such marks before.

On this occasion Tookta had only a couple of days' break, but during her stay Mae Jom managed to take her daughter for a walk down to the mulberry grove where they sat in the shade of a big, old tree. It was the first time since Tookta had been a young girl that they had been together like this. Both of them felt awkward. Both could remember the previous uncomfortable encounter with a clarity undimmed by the passing of time. Tookta had been fourteen when she had asked about her background and why she was a loog kreung. Mae Jom had never spoken to her about Johnny and the questions took her by surprise. But she was honest with her daughter, and took Johnny's photograph out of the drawer to show her. Tookta touched the image lightly with her forefinger.

'You took all this time, Mae,' she said with just a hint of bitterness in her voice. 'You kept him to yourself.'

'I'm sorry.'

'I look like him, don't I?'

'Yes, you do. Quite a lot.'

'And you didn't want to be reminded of him.'

'No.'

'Did you love him?'

Mae Jom had hesitated and then, trying to be honest, she had replied: 'I didn't have enough time to find out.'

'So you didn't have to love someone to go to bed with him and get pregnant.'

Mae Jom did not have an answer. Now, years later, it was Mae Jom's turn to ask the questions.

She began, gently: 'Do you want to tell me?'

Tookta looked away, shrugged and began to cry. Mae Jom wanted to reach out and take her in her arms and embrace her, instead, she held back.

'I'm sick. I've had it for a few years. I was an idiot. A farang offered me a big sum – a whole week's pay – to have sex with him without a condom. I should have known better. I did know better. What an idiot I was! But I've been taking the right medication. I'm looking after myself. Anyway, nobody dies of AIDS.' Tookta wiped at her face with a handkerchief. 'But that's not why I'm crying.'

'Well, what then?'

'I'm nearly thirty-eight.'

'But ...'

'Come on. You know that I've never been a secretary. I wanted to be in films but ended up as a hostess in a club. You must have guessed that?' She turned to Mae Jom. 'It was fine at the beginning. I was earning good money. But then, of course, they always want younger girls. Men want young flesh. So I got a job in a massage parlour, sitting in one of those booths. That was OK too; the money was enough so long as the customers chose you. Later, I worked as a foot masseur. But the place closed down. After the political troubles a lot of places went bust.'

'You can always come back here, you know. We'll take care of you.'

Tookta looked angry.

'I can't. It's impossible. I've been away too long. I've changed too much. I'm a city person now. I'm used to

Bangkok. I couldn't take it here in the countryside. Not for long. Not any more. It's too quiet. I'd go out of my mind.'

Barely a year had passed when Tookta telephoned Mae Jom to tell her she was moving into a temple that looked after the sick because she could no longer pay her rent and felt too weak to be on her own. She insisted that she did not want Mae Jom there, that she didn't need any help other than what the temple had to offer, that she preferred to be alone. She was calling, Tookta said, merely to say goodbye because she would not be visiting them again. There was little emotion in her voice.

After the call Mae Jom stood thinking for a long time in the field behind their house. It was dark and moonless. The stars were bright. She felt waves of emotion rising and falling in her chest. When she crept back into the house and into bed she could not sleep because the same thoughts kept playing over and over in her mind: was Tookta calling for help, or reproaching her for failing as a mother, or both? Mae Jom found it impossible to tell.

The hospice is part of a monastery in the Phrakanong district that stands by one of the city's larger canals. Nai Pot drops her off at the mouth of a warren of narrow lanes, waves to her, and weaves his pickup back into the traffic on his way to work. Mae Jom asks a passerby who immediately points the way; the monastery is a landmark and one of the few places where those who are poor and have the virus can find shelter and support. Mae Jom recalls seeing the abbot interviewed on television. He had kindness in his eyes.

Rows of taxis and tuk-tuks are parked at the entrance next to food stalls and shacks selling incense and flowers and other offerings. Mae Jom wonders at the crowds of people arriving and leaving and makes her way through them to the main temple. She does not know what she will say to Tookta, what words of comfort she might offer. She wonders if Tookta will agree to see her. She stops one of the volunteers who points to the women's quarters. When she mentions her daughter's name, the girl says: 'Oh yes, I know her, the loog kreung with the beautiful green eyes. She must have been a real stunner once.'

Mae Jom finds Tookta sitting on a bench under a neem tree shading a stretch of grass along the canal. She is reading a magazine and looks up at her mother as if she had been expecting her. Mae Jom tries to hide the shock of seeing her daughter's gaunt face and the brown patches covering her stick-thin forearms. She has lost so much weight that her bones jut out of the smock she is wearing. Mae Jom fumbles for words and, after a clumsy greeting, sits beside Tookta on the bench and stares at the water. It is one of the cleaner stretches of the canal, and on the opposite bank a young man sits on a small stool, a homemade bamboo fishing rod in his hands. The sunlight beats down. Tookta has a coughing fit and Mae Jom reaches out to hold her, half afraid this skeletal frame that is her daughter will shake itself to bits.

'It's the TB,' says Tookta, gasping for breath. 'I told you. Nobody dies from AIDS these days. There are good medicines. But the tuberculosis gets you every time; that or the pneumonia or the meningitis.' She lets out a rasping laugh.

Mae Jom nods, though she doesn't understand what she is talking about. Thinking that she has to say something and that now is the moment, she clears her throat and tells Tookta in one long breath that Yai Da is dying and wants to see her and they could take the afternoon train and go back to the village together. Tookta listens and smiles and reaches over to pat her mother's hand.

'Tell Yai Da I am too ill to go anywhere. She knows, doesn't she? Or haven't you told her?'

Mae Jom says nothing. Tookta shrugs.

'Well, it doesn't matter, Mae. Just tell her that I'm too weak to travel. That's the truth.' Tookta pauses, and laughs again. 'And tell her that we might be seeing each other again anyway, sooner than she thinks.'

That evening Mae Jom calls Nai Gawb on her mobile phone. She can sense that he misses her and that, despite himself, he is eager to hear about Bangkok.

'What's it like?' he asks. She detects in his voice a liveliness she has not heard for a long time.

'It's changed,' she says. She does not tell him that she barely recognizes the city, just as she barely recognized her daughter.

'Yes, but are there really so many tall buildings now?

They say it's as modern as any farang town.'

'Yes, there are lots of tall buildings, and trains go between them and under them.'

'Wow! What I'd give to see all of that. Did you go to Makkasan, where we used to live? Is there anybody still around that we knew?'

'No, I've not gone. Nobody will be there that we knew. It's been too long.'

She tells him that she has decided to stay for a few more days in the city to be with Tookta and he wishes he could be there with her.

'It must be really amazing to be back there,' he says. But she tells him no, she does not like what Bangkok has become, which she immediately regrets as it seems, at one stroke, to deflate his vicarious pleasure.

On her third day she decides look for the small hotel where she'd stayed with Johnny. She can't remember its name. Expecting the worst, Mae Jom feels a thrill as she rounds a corner and sees it, the three-storey converted shop-front. The noodle shop opening out onto the pavement, full of customers, foreigners and Thai, eating and drinking and talking is gone and the garish neon lights over the front door no longer offer cheap rates for long stays but promote a barber shop cum massage parlour – Thai, oil, foot and special body rub at discount rates. The shop-houses that once flanked it have been demolished and it is now wedged between two huge blocks: on one side, a bank, on the other a department store. Painted women dressed in tight black trousers and pink Polo tops sit around the entrance looking bored and tired, perched on low wooden chairs. They eye her suspiciously.

Through the glass front Mae Jom can see a few customers prone on massage beds, a blanket over the upper body, having their feet rubbed and their legs pummeled. The minds of the masseuses are far away. Mae Jom thinks of Tookta.

She looks up at another window, half-blocked by the dripping back of a rusting air conditioner. She recalls standing at the very window with Johnny. It was the evening before she'd left for Udon on the train. He was holding her from behind, arms around her waist, kissing her neck, all the while telling her that he loved her.

The next morning, on arriving at the temple Mae Jom is intercepted by the volunteer who has been looking after Tookta. She gestures her towards a quiet corner by a huge tamarind tree where she tells her in a soft, trembling voice that Tookta is dead. She says there is nothing to worry about, the temple will take care of the cremation; everything is paid for by a charity that funds the hospice. Mae Jom has expected such news and is glad she has heard it here, rather than read it in a letter. Nevertheless, she is stunned and, overcome by a moment of dizziness, holds on tightly to the girl's arm.

The poor who die of AIDS-related illnesses are cremated as soon as possible and that afternoon, in the crematorium at the back of the temple, they burn Tookta's skeletal corpse encased in a plywood coffin.

Three monks seated on a bench by the furnace chant a prayer while Mae Jom and Nai Pot and the girl volunteer stand by. In the courtyard they watch a plume of smoke rise into a cloudless sky; Tookta's broken body returning to the elements. Mae Jom thanks the girl for caring so much for her daughter. In turn the volunteer tells her that she has arranged for the charity to pay for Mae Jom's ticket back to Udon. It is an act of kindness she has not expected and tears well up in her eyes.

As the bus pulls out of the depot, Mae Jom again feels uneasy. She cannot explain why. Perhaps it is the news going around that the rains will be bad this year, and that there will be heavy flooding everywhere including the capital. The thought of being stuck in Bangkok without the means of leaving seems terrible to her. The city frightens her more now than it ever did.

II

The mysterious case of Khun Somphop

'We Thais are a peace-loving nation. But let no foreigner mistake this for weakness. We are ready to fight, to fight to the last, to defend our kingdom from any aggression, be it covert or overt, political or military ...'

Nai Pot switched off the radio as the cheering began. His boss was still asleep on the back seat, but had Khun Sompop woken and not heard the sound of his own voice, he might have been annoyed.

'They're going to broadcast the speech I gave last night in parliament,' Khun Sompop had told him. 'They said at the office that it went down well. I want to hear what the commentators think.'

Nai Pot had turned on the radio and tuned himself out as the speech began. He had heard it all before, many times, though this one was more hectoring in tone – more harangue than speech. It was part of the game, this need to show muscle. To the public, Khun Sompop Wongpanich was a future leader who could bring them prosperity and security and progress without compromising Thai identity; to his family and those who worked for him he was polite and considerate and kind; to the politicians, he was a player driven by ruthless ambition, merciless in battle, uncompromising despite his avowed commitment to principles of consensus and to bringing everyone together in a united front, one to be allied with or deeply feared.

Khun Sompop looked tired even as he slept. Nai Pot was careful on corners and skillful with the traffic until the dark blue Mercedes was clear of Bangkok's congestion and cruising along the highway heading north to the hinterland. Nai Pot glanced at him from time to time in the rear mirror but mostly focused on the unpredictable road surface ahead, given its tendency to pothole during heavy storms due to the poor quality of materials used in its construction – one of Khun Sompop's companies – and concentrated on not letting the suspension reveal its shoddiness. An hour or so into their journey, he saw his boss wake, disorientated, brain scrambling for a point of reference in the white light of the late afternoon. He wore the puzzled, frightened look of a man who might have been kidnapped. There was no traffic. The empty highway confused him.

'Are you OK, Sir? Do you want to stop somewhere?' asked Nai Pot.

Sompop sat up straight and rubbed his eyes, embarrassed. He had just been dreaming of being a little boy again, going to school in his father's car, worried that his shoelaces were not properly tied. Awake, he remembered now that he was on his way to a meeting, probably the most important one in his career.

'Umm ... no ... thank you ... no need for that,' he said, reaching for a bottle of water from the small fridge in back of the car. 'Where are we? How long have we got to go?'

'We'll be in Khao Yai soon, Sir. It's not far now. Look, you can see the mountains in the distance. From there, it's about a half-hour's drive up to the house, or at least that's what I've been told.'

'Did you hear the speech?' asked Sompop.

'Yes, Sir, I did.'

'And how was it? What did the commentators say?'

'It was a good speech, Sir. Everyone liked it.' Nai Pot knew what his boss wanted to hear; no details, just whether it was good for his image. 'Do you want me to put on the radio again?'

'No,' Sompop said, settling back in his seat, loosening his tie and unbuttoning his collar. He took a sip of water. His momentary disorientation worried him. It was not the first time. He had begun waking in the middle of the night with the same sense of confusion. He believed it was the work and pressure; another round of elections was expected to be called soon and he needed to be alert and in control at all times, asleep and awake, with the result that his mind was continually on edge.

Though he wanted to hear his speech, the particularly fierce heat of the day, which seemed to penetrate even the air conditioning and the tinted windows, and the two glasses of Chardonnay for lunch at the Italian restaurant, conspired to close his eyes even before they had reached the toll booth.

Lunch was with Anne, his Thai/American mistress; it was an unexpectedly intense and unpleasant encounter. Playing with her pasta and truffles and with a voice he considered uncharacteristically whining, Anne told him she needed more of his time. Her performance seemed well rehearsed, and caught him unaware. Sompop didn't like surprises and found her behaviour suddenly strange and disturbing. He had always thought their year-long affair to be what she'd wanted – carefree and independent – and their meetings, in a discreet, exclusive apartment overlooking the river, a welcome diversion from her otherwise hectic life. Anne was a much sought-after model and said when they'd met that what she wanted was a no-strings-attached relationship. She had no time, or desire, to play house. Both Europe and China were keen on her now.

This was fine with Sompop, who had his family-man image to protect and his own hectic schedule taking up his time. They saw each other when they could and enjoyed their intimacy without either staking claims on the other. They were adults. So why this sudden demand for more? Sompop sat with his handsome face a practiced mask that didn't quite hide his impatience, staring at her immaculate features, all the while thinking that if he did not desire her lithe, tender body with such an all consuming passion he would just fold his napkin, get up and leave without looking back.

'I'm in love with you,' she said, breaking their rule of not making public displays by grasping his hand.

Sompop glanced at the waiters, but they were engaged in other gossip.

'Can't you see that I want to be with you? Why are you being so cold? Why are you pushing me away? I can't stand it any more.' Her eyes were wide like a cat's and in the next instant, her head hanging down, she was crying.

Sompop flushed with annoyance but quickly recovered. He found the right soothing words, and promised they would have time together at an ultra-exclusive Chiang Mai resort. It seemed to be enough, though her body language said the conversation was far from over, which served only to annoy Sompop just when he thought he'd brokered a compromise. In the car, he was angry at himself for not having seen it coming. Maybe the affair had been a mistake from the start. He had a beautiful wife and two lovely children. He had never been unfaithful until he'd met Anne; what was he looking for? He was greedy, had dropped his guard and let himself be led by desire. Now he was paying the price, but he would find a way out. He always did. There was a lot at stake. He asked Nai Pot to turn on the radio so he could hear the reaction to his speech but promptly fell asleep.

It had been months since Sompop had ventured out of Bangkok and he found it a welcome change of scenery – no crowded pavements, no traffic jams, no smog. He'd told no one in his office where he was going, or what he was doing, just that he would be unreachable for the rest of the day. His staff, dependent on him for every decision, every approval, were not pleased with the subterfuge.

He felt a wave of release as the car sped along. He was no lover of the countryside, unless one counted the couple of rounds of golf he played each week at a club just beyond the urban sprawl. But then, along that stretch of the highway, there wasn't much countryside left anyway. Every hundred yards or so there was a new cluster of factory buildings and, after a patch of wasteland strewn with debris, came rows of pristine town houses, half of them empty, and a section of cheaply built low-rises. Sompop calculated that developers had a few miles left before the commute to the city outweighed the benefit of distance from the capital. It was a businessman's instinct, knowing just how far you could push before something became more trouble than it was worth. It was a useful skill for politics, too.

Thailand's progress excited him. He wanted to see a Thailand that wasn't viewed as a perpetual 'developing nation', backward and poor, unable to fend off the depredations of foreign companies and their governments, and denied the liberty to decide its own policies. That a country like Vietnam, ravaged by colonialism and decades of war could rise to become one of the most successful economies in Southeast Asia, stuck in his throat. With the right management Thailand, not Vietnam, could and should be a force to reckon with in a region increasingly falling under the economic influence of China. The time for pandering to US interests was over. His Thailand was no vassal.

Sompop prided himself on not being part of the political elite bound by favours that stretched back generations. He had come seven years earlier from the board of the family's property development company, Siam Land, one of the top three in Thailand. It had been a small construction company set up by his grandfather, who'd come from one of the country's old aristocratic families and trained as an engineer in England. He'd been in the hotel business during the Vietnam War and, anticipating the departure of the free-spending American troops, branched out into construction and development. In the mid-1970s, Siam Land was buying up and building on as much cheap land as it could grab heading east towards Pattaya.

His initial pitch was cheap, affordable housing. Siam Land banked enormous profit and was ready for Bangkok's building boom in the 1980s, raising the city's first tower blocks and changing the skyline forever. Sompop went to the best English schools and universities, studied economics, and could have written his own ticket, but chose instead to go into the family business where, for the next two decades, he embraced the only lesson his grandfather taught him – work hard, play hard – and took Siam Land to a position of even greater power in Southeast Asia's development boom. The company's logo was on many a joint-venture project in the early 1990s and dotted the Bangkok skyline.

At the age of forty-two, Sompop decided to expand into politics. He had for years watched from the sidelines with rising frustration and disdain at the failure of Thai leaders to apply the most basic solutions to the country's economic problems. He often joked, in private that few politicians would survive in the business world. What the country needed was firm leadership and a team that understood the workings of international finance.

The rise of Thaksin convinced Sompop that his time had come. The policies of the Thai Rak Thai party were a hotch-potch mix of populism and monopolies that would never make Thailand competitive in a globalizing world. In Sompop's eyes Thaksin was too provincial at home and too weak abroad, vacillating between courting the Chinese and cultivating long-established ties with the Americans. To him this was a confused foreign policy-realpolitik that was both unacceptable and unprincipled. Motivated by a sincere and aggressive desire to see his country join the ranks of Taiwan and Korea as modern manufacturing societies, as well as his own ambition that some saw as the consequence of an unchecked ego, he believed himself to be the only one capable of leading Thailand to its rightful position in the world.

Everything was in his favour. He was one of Thailand's richest men, had connections unmatched in the business community, and was fluent in Mandarin and English. He was intelligent, diligent, cultured and had film- star looks. He could be ruthless when necessary, but was sincere in his conviction that leaders should be held to high ethical and moral standards – something he'd picked up from his British education. Sompop rose rapidly through the ranks of the now defunct, extreme right Righteous Nation, which put him in a strong position to cross to the more attuned True Thai party. His trim, athletic figure contrasted with the bloated, cynical politicians who sat around him in the parliament, and his quick mind easily made mincemeat of opponents in debate. He became the poster boy of political commentators, who declared he had the charisma of a natural leader.

Even those who resented his arrogant, self-assured manner and were jealous of his privileged background had to nod begrudgingly to his persuasive power and to concede, on analysis, that much of what he said made sense. Here was a real alternative to Thaksin, a bright light at the end of a very dark tunnel.

But Sompop knew it was dangerous to move too fast in Thai politics, and while Thai Rak Thai held the power he bided his time, strengthening his alliances, waiting for the right time to strike. He established and tended his power base and learned to ride the merry-go-round, climbing step by step up the ladder of power. His political astuteness and survival instinct helped him weather the political upheavals and scandals through three successive governments, as many other veteran politicians slipped and fell into the abyss of public condemnation or, worse, total oblivion. Under Thaksin there was much finger pointing at politicians who exceeded even the Thais' very liberal interpretation of excess. In the end, Thaksin fell victim to his own campaign and was chased from office. Sompop emerged untainted, rich enough to be above financial corruption, though he deliberately avoided making a show of this. If anything, he dimmed the light of his charisma rather than parade his capabilities. In doing so, he made the public want even more of him. Suddenly, and the pundits were at a loss to explain just when it happened, everyone agreed he was a major contender for the biggest prize.

By Sompop's calculations, the elite would turn against Thaksin, self-styled champion of the poor, sooner or later, and after the military coup of 2006 he saw that he might, sooner rather than later, have to take his shot, but not yet. He distanced himself from the brief fiasco that followed a period of incompetent military government, and he demurred about joining forces with the Democrat Party. He wanted to become prime minister through a real election and to lead a party that won not only the people's confidence, but their hearts and minds. He was now forty-nine, no longer dismissed as a 'rising star'. The name on people's lips was Sompop Wongpanich. Only he could fix Thailand's problems.

Wild natural greenery took over from where the developers had stopped, something that registered on Sompop only because the light seemed less harsh now, and the car began the gentle climb into the Khao Yai district. It was a long way to go for dinner, but the man he was about to meet was the billionaire head of Austro/Thai Holdings, Dr. Rawee, who was in a class of his own in Thai politics; his connections ran through all the factions of the power structure, embraced key military and business leaders, was said to reach deep into the palace, and held sway over the Buddhist north and the restive Muslim south. He shunned the spotlight and maintained an image of lofty disinterest, but in the constantly changing landscape of Thai politics, he was recognised as a key figure who could make or break you. His hand had been suspected to have been at work several times in the power struggles of the past decade. Sompop had met Rawee only once, during a formal event at the U.S Embassy, but the older man had shown no particular interest in him other than mentioning that he had been senior to Sompop's father at the same boarding school in England. He was known for his eccentricities, which included cloistering himself in his well-guarded compound. The was why Sompop could not let slip the chance of meeting him now, at this crucial stage in his career. He felt instinctively that his future was riding on this face to face exchange. He was confident Rawee would support his vision for Thailand.

The audience with the great man was agreed to take place out of town away from the prying eyes of the press and the media. This was inconvenient but to be expected. But it was the mutual acquaintance who had brokered the meeting that bothered him, and had he been less ambitious and sure of his instincts he might have been more suspicious. For the broker was none other than Sia Oui, who had retired from politics when members of the party he was poised to lead to electoral victory, or at least to a position of enormous influence in the next government, jumped ship to join a rival coalition. Sia Oui hadn't had a hint that he would be betrayed.

It was Sompop who had engineered the coup against Sia Oui. At the time he had hoped that Sia Oui would see the move for what it was – politics – and accept that, in this, the student had outdone the master, for it was Sia Oui who had taught him how to navigate the crocodile-infested waters of Thai politics and introduced him to the people who most mattered: the largely anonymous faces that lurk in the shadows and decide which policies are adopted and which discarded.

Sompop never acknowledged any debt to Sia Oui, but readily admitted his early career had benefited from the advice of such an eminent politician. Sompop was beholden to no one.

Still, to cleanse himself of his feelings of responsibility for Sia Oui's subsequent downfall, Sompop made a generous donation to the Buddhist University in Ayudhya, which had then built a new wing in Sia Oui's name. But Sompop's attempts at making merit – a respectable Buddhist practice that allows the waters of guilt or innocence to be muddied – did not ease his discomfort. Time, though, had taken care of that, and Sompop managed to push the potential karmic consequences of his actions deep into the recesses of his subconscious, where they had remained until now.

Sia Oui's telephone call to tell him Rawee wanted to meet had unsettled Sompop, but he took it as an act of conciliation, even political respect, that his former mentor should involve himself again by bringing him face to face with the one man who could offer the golden chalice. Had Sia Oui harbored ill will, he might have simply refused to act as intermediary. Sompop was ready to accommodate any demand the old man might have and, in doing so, finally expunge his karmic deficit. And besides, he would need someone like Sia Oui, whose only interest was self-interest, which was best served by doing the bidding of the master he served. He was a fixer par excellence; Sompop shuddered to think what Sia Oui might have done in the past.

'I think we'll be there in about five minutes, Sir,' Nai Pot said as he turned the car off the main road and onto an immaculate stretch of tarmac.

Sompop grunted acknowledgement and began to focus on the meeting ahead, the possible questions that would be put to him, how he would answer them. He had done his homework. He had a good idea what Rawee wanted from him and knew where his interests lay. There were several topics on the agenda. They would be concentrating on the border troubles with Burma, which threatened to escalate into a full-scale conflict, much to the annoyance of many in Thailand who had business dealings with the Myanmar regime. They would certainly be discussing the economy and how to revive it, and the Islamic insurgency in the south, which was now reaching a critical juncture. He had in mind various solutions and was sure Rawee would see in them a mixture of pragmatism, expediency and common sense. In his heart, Sompop was already celebrating the outcome of this evening.

Sia Oui liked the quiet twilight, void of the noise of so many birds and animals and insects. The jungle had its own language. At twilight, though, during that brief half hour when nature revealed itself with the clarity that artists sought and failed and sought again to capture, everything was still. Before the beautiful spectacle Sia Oui was sometimes moved to awe. But that evening, as he stood by the wooden railing watching the golden glow in the sky fading into deep purple, he was uncertain of his feelings. On waking that morning he had experienced a warm rush of excitement he had not felt in a long time, the feeling that he was back in command, but now he was nervous. Everything had to be done the way he had planned so meticulously over the past weeks.

He stepped onto a small platform jutting out from the covered teak walkway, unhooked a pair of metal tongs and, from a plastic bag he was carrying, drew out a large piece of raw ostrich meat, which he flung into the water. He heard the splash and seconds later there came the sound of violent thrashing as the reptiles fought for the food. He hurled another piece towards the sound and another and the thrashing became more violent. Sia Oui often remarked to visitors that the crocodiles reminded him of the politicians in Bangkok. But unlike him, few of them had the stomach to deal with the real, dark and distasteful side of Thai politics.

A few feet behind Sia Oui stood Toi and Ti, ide[illegible] twins. They were handsome, fashionably dressed, fit, dark-skinned young men in their mid-twenties. He had found them on the streets as thin, grubby urchins and brought them into his home. They were now his bodyguards, his secretaries, his sexual companions. It was common knowledge their loyalty to him knew no limits.

'Ai Toi,' said Sia Oui over his shoulder as he flung the last of the meat into the water. 'Is everything ready?'

'Yes, Sir. It is all set up.'

'We have taken care of every detail,' added Ai Ti.

'Good. Then we'll have some fun tonight,' Sia Oui grunted.

Half an hour later, freshly bathed and dressed for the evening, he was sitting on the balcony, a whisky and soda on the table beside him, jazz drifting in the air around him as he enjoyed his first drink of the day. Before him stretched his vast estate sloping down to the valley and up the thick, green hill that joined the others undulating into the distance. It was idyllic, and it was his alone to see enjoy. He had promised his parents before they died that he would make them proud. He would not be sucked under now, not after they had slaved away on their smallholding in Lopburi and at whatever cash work they could find to keep him fed and clothed and educated.

He was listening to Duke Ellington – the 1943 At Carnegie Hall – his favourite recording, the rich and complicated chords. The nervousness he had felt earlier was gone. In its place he felt calm tinged with melancholy. He picked up his glass and twirled it around to enjoy the clear clink of ice against crystal.

It was almost exactly three years since he had last seen Sompop. That meeting had taken place in the office he had occupied for the whole of the time he worked for the Righteous Nation Party. It was a glassed off corner of a large open floor, and he could watch everything from there. It had been an intense exchange. Party members and office staff had walked past, pretending not to notice, and both men were close to tears as they said goodbye.

Sompop triggered in him contradictory emotions. He came to think of Sompop not so much as a subject of physical desire, though there was that, but as a son he had groomed for success, and such was his brilliant promise that the top job was not out of the question. He had great plans for his protégé. And then the betrayal, the more vicious for being so sudden.

He took a long sip of his whisky.

Sia Oui's bid for the party leadership had been coming to a climax. The foreign press dubbed him 'Mr Fixit' for his outstanding ability to support, cajole, threaten and eliminate members of Righteous Nation. Sia Oui had worked hard to prepare his party to lead a broad coalition that would be a viable opposition, and ultimately an effective government. The lynchpin of the coalition was Righteous Nation and, thanks to his carefully drawn strategy, the future of Thailand would soon be in its hands. The Burmese astrologer who had served his late wife and stayed on, at a price, to steer him through his political journey had told him this. His charts and calculations revealed a brilliant future. Granted, there would be a difficult period at the beginning of the coalition, but that would pass. There was every reason to be optimistic about the future. Sia Oui would gain complete control of policy making. There was no doubt about it. The fact that he had failed to predict the truth meant that his advice was never sought again by the rich and powerful.

Without the slightest warning, Sia Oui faced a mass defection from Righteous Nation. Even those whose loyalty both to himself and to the party he had taken for granted were gone. A brief statement, signed by the defectors, explained that, due to their insoluble differences with the policies of Righteous Nation, they felt they could no longer remain in the party without compromising their integrity. Sia Oui knew that in some dark place out of sight of his vast network of informants, the conspirators must have worked for months to mount their coup. The final twist of the knife was that they all went to True Thai, a relatively new party of politicians with whom Sia Oui had less than amicable relations He had lost face in the worst possible way.

The media had no clue what questions to ask and Sia Oui was able to brush them off with answers that sounded confident and dismissive but when transcribed said nothing at all and so reporters were left to speculate, often wildly, and within a week or two the story had dropped to page four. Sia Oui took consolation from

Sompop's loyalty; he'd stuck by his side as the situation went from bad to worse, and then hopeless. There was nothing left of Righteous Nation. Sia Oui announced he was quitting politics, and quietly advised Sompop to distance himself from the disaster and continue his climb on a more stable ladder. Sompop, with apologetic bows, said the most pragmatic choice would be True Thai – he already had allies there in his former colleagues. Sai Oui could only agree.

Khao Yai, an occasional retreat, now became his home. He had arrived there long before the wealthy discovered the hillsides to be a convenient getaway from Bangkok and built their mansions and condos. His was the prime spot and his estate was vast, a whole mountainside. He had invested a great deal of his wealth into it; the grand mansion that was a replica of the White House. As a farmer's son, he believed the land was there to be worked, and he built farms and orchards and planted them so fresh produce was always available. It was virgin soil uncontaminated by development, and he saw early the market potential for organic produce in the West. Khao Yai had been conceived as a plaything, but even before leaving Bangkok it was something bigger, and in profit. In the first months of exile, however, all he could do was sit for long hours on the balcony and brood and pick at the scab of his disgrace.

That he had been betrayed was obvious; by whom was not. The timing of the operation was too perfect, and promises would have been needed, which meant someone in his own party had collaborated, if not help to orchestrate it. He did not want to embarrass either himself or others by using his personal connections so soon after his humiliation, so he sent the twins to interview one of his former underlings, a man high on his list of suspects.

'Have as much fun as you like, boys,' he told them. 'But don't make too much noise about it.'

Ai Ti and Ai Toi were back within twenty-four hours. The man was not the ringleader, and it took disappointingly little to persuade him to name who it was, and furnish proof. Regrettably, the man was depressed, with personal problems, and evidently so suicidal he had stuck a syringe containing rather too generous a cocktail of heroin and cocaine in his neck.

The name, and the proof that bore it out, filled Sia Oui with bitterness.

As the Mercedes was waved through the tall iron gates of the estate, Sompop buttoned his collar, straightened his tie and checked his hair. He touched a box at his side. It contained a bottle of vintage Krug, requested by Sia Oui; it was Dr Rawee's preferred beverage.

'We can look forward to a long night if the old man is in the mood,' Sia Oui had told him over the phone.

The long curving driveway took Sompop past an enormous compound of ostriches and the crocodile farm, across which ran a well-lit covered walkway. Sompop didn't see any crocs, but he had heard that Sia Oui's crocodile skins graced the arms and protected the feet of the most affluent and elegant women in

Europe. The car's halogens swept across a neatly planted vineyard, orchards, vegetable gardens.

'He's returned to his farming roots,' colleagues would joke in Bangkok. 'He'll be starting a green party soon.'

Sia Oui waited in the portico, and clapped his hands like a young boy as the limousine slowed. He looked trim and fit as he bounded down the front steps of his mansion to the car, even before it had come to a complete halt. He was smiling broadly and took Sompop's hand as he emerged and bowed politely. He hugged the younger man, giving him a comradely slap on the back.

Sompop was relieved by the welcome.

'Well, well. It's good to see you again. It really is. But you're looking tired. Was it a long ride? The traffic? You should get out of the city more often. Politics getting you down?' Sia Oui was in fine form, laughing, relaxed. 'Come! Take off your tie. Get comfortable. You're not in Bangkok now.'

Sompop was struck not just by the warmth of Sia Oui's welcome but by his transformation. The last time he had seen him he'd looked frail and tense, trapped in his defeat by a crowd of media pushing recording machines into his face and blinding him with flash guns and camera lights.

Sompop hid his surprise behind a charming smile and in his own greetings complimented the older man on how invigorating the fresh country air must be. He commended Sia Oui for his bubbling, positive energy.

He signalled to Nai Pot that he would call down when he was ready to leave and allowed himself to be led through tall, magnificently carved teak doors into the house. Sompop made appropriate sounds of being impressed. They agreed that it was too late for a tour of the grounds but Sia Oui led him at a slow enough pace down marbled hallways past open doors and glass collection cases of ancient pottery and rare museum-quality artifacts and walls of paintings and tapestries to ensure Sompop got the full measure of his wealth and fine taste.

At the far end of the house they went through French doors that led them out onto a sweeping balcony overlooking the gardens, where ground lights suddenly came on to reveal a breathtaking tableau of botanical wonders. Sia Oui was delighted when Sompop audibly gasped, and took a shoulder and led him towards a conversational arrangement of comfortable chairs. Ai Ti stepped from the shadows and placed drinks before them: two fingers of smoky whisky in heavy Baccarat glasses.

Their conversation was polite and impersonal. Sia Oui asked Sompop about the health of his father, whom he had met many years earlier, and the family's construction business, and about his wife and children. They talked about the World Cup – would Thailand ever be strong enough in football to host such an event? Probably not, they laughed – and of other events that neither concerned nor particularly interested either of them. Sompop heard how well the vines had taken to the climate and how the grapes were already producing small quantities of a promising wine, but it had only been five years and one mustn't rush these things, and the ostriches were now appearing on the menus of several of Bangkok's finest restaurants, and his organic gardens were proving, as he had expected, to be a model worthy of emulation across the country.

'The market for certified organic produce is staggering,' he said with a laugh. 'All the farangs who have money insist on eating only organic.'

He reserved his best hyperbole for his crocodiles, and talked at some length about their proper care and feeding and concluded by saying, with no attempt at modesty, that he was getting calls from the leading French fashion houses for his skins. Khao Yai was going to be at the center of a farming revolution, he said, one that would raise Thailand's farmers to undreamed of success and, of course, make him even richer than he already was. This last was no immodest boast, Sompop understood, merely an honest statement of fact. Sompop kept up his part of the exchange with ease, politely agreeing with Sia Oui and showing his interest in everything he was being told. He encouraged Sia Oui to expand on the joys of country living. At no point was politics, or the tardiness of Rawee, touched on, even by allusion.

'No,' said Sia Oui, sipping at a second drink, 'I could never go back to Bangkok. It's become such a contaminated city. It's rotten through and through and, I fear, on the brink of collapse. I'm sure of it. I love the life out here.'

Somewhere in the house a telephone jingled and shortly Ai Ti emerged with a handset and leaned down to whisper in his master's ear. Sia Oui nodded and excused himself – 'I must take this' – to the far end of the balcony. Sompop watched what appeared to be a deep conversation. He guessed, rightly, who the caller was, for Sia Oui wore an apologetic smile on his return: 'Dr Rawee has been delayed. But he is on his way. He said we were to start dinner without him.'

Sompop was unable to hide his disappointment; without the informality that preceded dinner, he would not be able to properly tune his pitch. Sia Oui showed sympathy.

'Be patient. You've come all this way. Relax a bit. He'll be here soon. He is a man of his word and you and he have a lot to talk about, don't you? If the night goes too late, you are welcome to stay.'

Sompop politely declined the invitation with an excuse an early meeting in the morning at his office. In fact, he had been hoping to call on Anne before getting home; one of his placating promises over lunch.

'As you prefer,' said his host.

It was dark, the sky was clear, and the stars were sparkling like diamonds scattered across velvet. They sat, listening to nature's nocturnal sounds.

Presently, Ai Toi appeared and announced that dinner was ready.

'I always dine early here,' Sia Oui said as he stood and led the way into the dining room. 'It's much better for one's health.'

There were three settings of silver and crystal and linen on the broad round marble table. A fan turned overhead. Beside each place a red candle standing in an elaborate gilt candlestick had been lit and around the walls there were other candles. No electric light. Sompop thought it too theatrical.

They sat opposite each other, leaving Rawee the setting between them for when he arrived. There were no servants, only Ai Toi in the corner awaiting his master's orders and no sign of Ai Ti. Sompop's face must have betrayed something of his thoughts because Sia Oui said: 'There's a fair at the temple in the next village tonight and I've let all the servants go out and enjoy themselves. There's not too much nightlife here in the hills. We have to learn to entertain ourselves.' He laughed.

Sompop laughed too, and hoped his relief didn't show.

Sia Oui signalled and Ai Toi was at Sompop's side with a bottle of red wine, label on display, and Sompop raised his eyebrows.

'One of ours,' said Sia Oui. 'Let us put this to the test. It will be a few years before we can engage in serious production, but personally I think it's better than the French. It's quite dry and goes well with the ostrich stew.'

Ai Toi opened the bottle with practiced skill, and poured it into Sompop's glass. Sia Oui waved him away. 'I'll stick to my whisky for the moment,' he said and raised his tumbler towards Sompop.

'Cheers! To your successful career!'

Returning Sia Oui's smile, Sompop, who liked wine, took a long sip from his glass, looked thoughtful, and declared the wine delicious. Sia Oui nodded his head, graciously accepting the compliment. For a while they were like two old friends enjoying each other's company. But it seemed to Sompop that in the candlelight, which danced with the turning fan, Sia Oui had grown suddenly ancient, wizened, almost reptilian, like an old peasant worn out by a life's labouring in the fields under the hot sun. The lines of his face deepened, darkened, and the flickering light had the effect of making them appear to oscillate. Was Sia Oui about to cry?

The thought was both absurd and amusing to Sompop, and to cover his urge to laugh he took two generous sips of his wine and set down his glass. It was immediately refilled. Sia Oui fixed on him a sorrowful gaze before turning slightly, as if addressing someone else. His voice was slow, droning.

'I really am so glad you decided to come. I wasn't sure you would. But it must be a question of karma. You had no choice but to be here. Because I have something very important to tell you this evening that concerns both of us.'

Sompop strained to hear Sia Oui; his words seemed to come from a long way away and he wasn't sure he understood them.

'Can you guess what it is?'

Sompop was too surprised to reply. All he could think of was why Sia Oui was saying these things to him, and mentioning karma. In any case, the older man left him no time to say anything.

'Dr Rawee's support, this is what you are seeking tonight, isn't it? You think this meeting will be crucial to your political career. Am I right? Of course I am. And in spite of your doubts about accepting my invitation you had to take the risk, didn't you? The temptation was just too great.' Sia Oui let out a chuckle. 'Hell, I taught you everything you know, every trick in the book ... well, perhaps not every trick. You forget that I was the best political broker in the business. But you of all people know how great I was. You stand on my shoulders. You think you are about to lead the country out of the dark days and into the light. I heard the speech you gave last night. You've really come a long way. Sabre rattling is always good for a spike in the opinion polls. It shows them that you're not just clever, you're also macho. It's always a risk, though. You'd have to back it up, and I'm not sure you can.'

Sia Oui sipped his whisky. Sompop was silent, not yet sure what he was listening to, but too polite to interrupt his host.

'If I am not mistaken you know exactly what you are going to say to Dr Rawee to win him over to your vision. You have your message so well- rehearsed that the words will trip off your silver tongue. You'll seduce him and he will back you on your march to the top. You're eloquent and persuasive. That is your gift. The others ... they couldn't sell a fake Rolex to a farang. No, you're the smooth one.'

Sia Oui finished his drink and Ai Toi took the empty glass and replaced it within moments with a fresh drink. Sompop continued to sip at his wine. If he felt discomforted by what he was hearing, he gave nothing away, except to shift a little forward in his seat.

'Oh, my dear Sompop, you must be annoyed at me for talking like this. But I make no apologies. I feel that I have the right to say anything and do anything to you.

'I was so angry when I had found out what you had done to me. I hadn't been aware of your treachery because you did it with such skill. Sticking with me to the end was a nice touch. There were so many fingerprints on the knife that yours, the first, were obliterated. What a fine execution! What ruthlessness. To coldly cut down your friend and your protector!

'I have seen a lot of things, done a lot of things, in the name of politics. But never have I seen such disloyalty. But I shouldn't have been surprised. After all, this is politics. This is Thailand. "Amazing Thailand".'

Sompop decided he'd had enough but when he tried to speak his mouth would not form the words. And the harder he tried the more confused he became.

'Please don't insult me by justifying yourself. I'm not really interested in your apologies. What's the point? They'd only be lies. I'm not looking for an explanation. You're a pragmatist. You did what you did to further your position. That's just business. But your treachery hurt me very deeply. I was a monk when I was young and I tried meditation and prayer and delved deep within myself, but I could not find a way of letting go of the whole miserable episode. None of it helped me. I lose face, I lose everything. That's how it is with us, isn't it? We politicians are meant to have thicker skin than other people – crocodile skin. But for the instrument of my downfall to be a person I once loved, someone I considered to be a friend? So ironic! Perhaps you will taste that bitterness one day.'

A numbness began to spread through Sompop's body. He could not even move a finger but tried not to let his panic show.

'Ah,' continued Sia Oui after a pause, 'I see the wine was particularly good tonight. I hope you aren't too uncomfortable. Ai Toi prepared the syringe himself, you know – didn't you, Ai Toi? He took care not to damage the cork when he made the injection. He tells me the dose is not lethal – accidental death is not what I have in mind for you. What I do require is your undivided attention. I trust you can still take in what I am saying? I've been preparing for this for a long time, and I do so like a captive audience.'

Sia Oui smiled at his own bad joke. He wandered in and out of Sompop's sight line, sometimes close, sometimes further away. Sompop could also see Ai Toi, hands crossed in front of him, with his brother by his side in the same, guard-like stance.

'Shall I tell you what I have in mind for you? Hmm? I've thought about killing you with my own hands, wringing your handsome neck, squeezing so tight so that I might look into your eyes as your life came to an end.'

He snuffed out a candle between thumb and forefinger. Snuffed out another.

'But I was angry then. I have never killed anyone, and I don't want to start now. I have worked hard to accumulate merit. My karma is in balance. It would be better if I had the boys do it for me. I believe it is their destiny – and they do such nice work, very tidy, always impeccable. Your body will go through the impressive mincing machine we've recently imported from Germany – to process all our organic meat – ha! The crocodiles must be fed. Your flesh will make a change from the ostrich meat we give them every day. You, with your fine breeding and affluence, will be a delicacy for them. I doubt they will enjoy your driver quite so much, but I've not known them to be fussy eaters.

'Your beautiful car will be spared. I expect it will be driving the streets of Phnom Penh long after we have all turned to dust. There will be a search for you, of course. Your calls will be traced, and I will say I waited up for you because our reconciliation was long overdue, but you never arrived, never even called. How I thought you still bore me enmity and how relieved I was to know that you were on your way here. The police will investigate, because I will urge them to do so, and the media will be in a frenzy and the unwitting Anne will be discovered and suspected – what a scene she made at lunch, I couldn't have planned it better myself – but they will find nothing, no clue, and into their files will go the mysterious case of Khun Sompop. Better still, Siam Land shares will plunge on the news of your disappearance but, rest assured, I'll be a buyer when they bottom out. Such a good investment; it'll thrive with or without you. Maybe I should try for a takeover? Shouldn't be too hard to persuade the board to take it public with you gone, dilute the family stake, buy them out.'

Sia Oui began to laugh, not diabolically, but in a quiet, satisfied way that built slowly until it slipped from his control and echoed off the marble floor and bounced around the walls. It became too much for him, and a hacking cough took hold. He doubled over. Ai Ti rushed to his side, helped him back to his chair, raised the whisky tumbler to his lips. All this Sompop watched through eyes wide with terror.

Sia Oui recovered his breath.'Yes, I'm satisfied that this is the way I will punish you. It is primitive, I must admit, and totally crude. But then that is what revenge is all about, isn't it? There's nothing subtle about it. And you know you deserve it, don't you? You, with your fine principles, showed no mercy. You should have had me killed. Instead, I get to teach you one last lesson – what ruthlessness really means. Do you understand?'

Sia Oui sat back and raised his glass, which seemed to have grown heavy in his hand, as if to toast his guest. He gazed at the dregs, let out a long sigh and, without another word, put the glass down, rose quickly from the table and left the room. The twins went with him.

In his solitude, Sompop listened to the call of a night bird and the muted drumbeat floating up from the temple fair. He rode the wild river of thoughts that ran through his mind until he reached calmer waters. Sia Oui clearly wanted him to think of escaping, which he did, but elsewhere in his mind was the urgent desire to explain himself, justify his past actions, negotiate – Sia Oui was not a psychopath, he would listen to reason. But the messages he tried to send from his brain failed to reach their destination; he was paralyzed. His breathing was unaffected, and his thinking was clear. He would be alive and aware when he was fed to the crocodiles. His stomach was churning in acid. He thought of his precious children, his devoted wife, his neglect of them, and he thought of Anne and how he would never again touch her soft, sand-coloured skin and long silky hair. He wanted to scream out, to fight for his life. But he could not.

He heard footfalls come up behind him and a smiling Sia Oui and his twins were in front of him again. Ai Ti held in his hands a short, thick length of rope. Perhaps he would be strangled first. That, at least, would be merciful.

'I've let you meditate, to prepare yourself,' Sia Oui said. 'It is time to go.'

The twins, one on each side, took hold of Sompop and lifted him from the chair. His legs were like those of a rag doll's and the toes of his shoes left black drag marks on the marble as they went out through a side door to a parked van. He was lifted inside and propped up between the twins who only slightly loosened their grip. Sia Oui was in the driver's seat and started the engine.

Sompop wasn't a religious man, though like every other good Thai he paid lip service to the Buddha's teaching and went to the temple to light candles and joss sticks and make merit when the occasion required. He disliked how people used religion to avoid taking responsibility for their actions, blaming evil spirits for bad deeds and thanking the Buddha when things went right. Sompop believed in scientific rationalism and materialism, that life wasn't easy, and rarely fair. But in the back of the van, minutes from death, he prayed with the fervour of the condemned.

The van slowed to a halt, its tyres skidding in gravel as Sia Oui jerked on the parking brake.

One last supplication came to Sompop's mind: 'I will give up my ambitions – please, let me live.'

The twins were careful as they hoisted him up the steps to the walkway and carried him onto a small platform over the water. Sompop saw metal tongs hanging from a hook and flecks of blood on the teak railing and walkway planks: a feeding area. From under the walkway lights came on. In the water below he saw movement and heard splashing.

Sompop heard only the clank of metal on metal, and imagined a chain and winch. He prepared for his death. Absurdly, he worried that all the dragging had ruined his shoes and that his laces were untied.

When he awoke he was naked on a wide bed under a mosquito net. The high-pitched trill of a bird drifted in through the open window. Judging by the light, it was morning, still early. His breathing was heavy and laboured, his head throbbed and his whole body was covered in sweat; it smelled sour. He raised a trembling hand, looked at it closely, raised it slowly to touch his head, his forehead, his nose, to make sure he was not dreaming. He parted the mosquito net and sat on the edge of the bed. His clothing was neatly arranged on a valet, shirt pressed, shoes shined. A fresh towel, cotton slippers and a light robe had been placed on a rattan chair.

Sompop showered, dressed and took his phone from a low pocket inside his jacket. Nai Pot sounded bright and cheerful, answered that he had been well looked after and well fed, and responded that he had watched the football match on a big-screen TV and had slept in one of the staff quarters out near the gatehouse. The car was washed, cleaned, refuelled and ready. He would be at the front door in three minutes.

Usually after a conference or important meeting Sompop would talk to his driver about what had gone on, who had said what, whether the time had been well spent, or simply wasted. It was how he unwound after such public appearances. The whole encounter would be described in detail and he would invite Nai Pot to offer his comment. But on the drive back from Khao Yai, he said not a word.

It was not for Nai Pot to ask, but what puzzled him was that no other car had arrived at the estate after he'd parked the Mercedes. In fact, there were no other cars there. He had expected to have a few drinks with the other drivers, but he was alone all evening. He knew someone important was meant to be joining Khun Sompop and Sai Oui for dinner. When he'd asked one of the servants who had not gone to the fair about this, all he received was a shrug.

As the road levelled out and straightened at the foot of the mountain, Nai Pot glanced in the rearview mirror and saw his boss removing a white envelope from his jacket pocket. Khun Sompop turned it over and over in his fingers, as if searching for some clue as to its content. He looked confused. He slipped a finger under the pasted down flap and tore it open, extracting an expensive sheet of heavyweight cream-coloured stationery. His brow crumpled into a series of ripples.

Two days later, Sompop announced his retirement from politics. He wanted to take care of his family business, which was going through a challenging period of transition and restructuring, and politics would be a conflict of interests. No one believed a word of it. But no one could challenge him. Family comes first. It sounded noble enough and true to his principles. There were those disappointed that he had not put the good of the country first, and speculation as to why, when so close to the prize he would walk away from it all. Other politicians had managed to keep their business affairs at arms' length, though a few had found the temptation to legislate to their advantage too great to resist, but it was to his credit that he wanted to avoid even a hint of corruption. The pressure from politicians who had gambled everything on his success was intense. They demanded an explanation. They even threatened him. But Sompop was adamant. In private, he told his family he was disenchanted with politics and its compromises. They did not push him for further explanations.

With a new round of demonstrations breaking out up and down the country and military rule again looming like a specter, attention turned to national issues. If his name was mentioned again it was in the context of a debate over the firm, decisive civilian leadership that was needed in the country. Some commentators would then express their regret that Sompop Wongpanich was now out of the picture. His story was soon forgotten, even as the most important stories are. Only his ex-driver, Nai Pot kept the memory of that strange evening with Sia Oui alive in his mind, returning to it again and again, certain that, along with the letter he'd glimpsed, the evening held the key to Khun Sompop's decision.

III
Isaan Dreaming

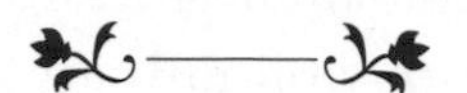

A Thai mansion, bordered on three sides by canals, beyond Onnut, over the bridge at Phrakanong, the countryside in the 1970s to those who lived in the old quarter of Bangkok, a place where pythons thirty feet long, or a hundred depending on the teller, appeared and disappeared, where water buffaloes ambled the lanes off the main road, deaf to car horns that only served to set the wild monkeys screaming from the tall trees; the canals busy with canoes going to and from the local market and the water clean enough for swimming, and children jumping in from the small bridges; the mansion's motorized wooden boat ferrying guests from the Phrakanong landing to the house on special occasions – such as Loy Kratong, the mid-autumn water festival, when everyone dressed in old finery and a traditional orchestra played on the lawn.

This was Nai Pot's childhood. In later years he would try, unsuccessfully to work out whether it was Chance, Fate, or Karma that had led him and his mother there. Nai Pot was two when Mae Lien left their village in Udon in 1970 when her husband, a mercenary in the covert war the CIA were conducting in Laos against the communists, was killed in the jungle. She could have left her son in the care of a granny in the village. This would have been normal. Other women had gone to the capital for work, sending cash home when they could in the hope of securing a future for themselves. They said it would be easier if she went alone. But Mae Lien could not bear to be parted from her baby and insisted she would manage, despite the obstacle of being a single mother in the city.

On the long train ride to Bangkok, she shared around the food she had prepared for their journey with the other women on board and got talking to one of them who worked as a maid in the Wongpanich household and who told her they were looking for an assistant cook. Confident that the angels were with her, Mae Lien, with her one battered suitcase and her two-year-old boy, took the bus with this woman to the Wongpanich mansion, was accepted for the job and gave thanks to the Buddha for her blessings. It was as simple as that.

Because he was hardly aware of where he was when they became part of the Wongpanich household, and not having known any other family Nai Pot never felt he was out of place there. And although from the beginning he understood his position in this idyllic world as that of a servant's son, he was never ashamed or self-conscious, nor was he made to feel so. He played at the back of the house, floating paper boats in the canal, or splashing in the water while his mother prepared the meals.

Sometimes it would be simple fare, other times on a banquet scale, and as he grew older he learned to help her and to serve at the table. He became skilled in etiquette, so much so that to him it seemed normal there were some places he could not go, some things he should not say, things he could not do. There were complex rules of language that applied in and around the house, one for the family, another for guests, another among the servants according to their seniority and duties. When his mother became head cook, he noticed a change in his own status, and the language around him.

The family sent him to the local school where he became gradually aware that he was living in a kind of bubble. No one believed him when he talked about the Wongpanich household. They laughed and said he was making it up.

'Nai Pot, where you live is like one of those soap operas on television,' a friend laughed when they were in their early teens. 'You know, all those Khunyings and Khunais sitting around waiting to be served.'

Such comments stung but briefly. He and his mother were more comfortable than the most of his friends' families and they were treated well. Besides he knew no other life than the one in the mansion.

The head of the clan, Khun Daeng Wongpanich was a formidable patriarch, a widow who had retired from the day-to-day operations of the business he had built but who was still in command of its affairs, and those of his family. He was treated with great respect, not only by the immediate family members living under his vast roof, but also by his guests. He had once been a well-known figure in Thai society. But after his retirement he no longer took part in the social life of Bangkok despite the endless invitations he received. He mostly stayed at home and busied himself with his orchids, blooming perpetually in a shady corner of the large garden. When he tired of being outdoors, he would go up to his studio and paint dreamlike scenes from the Ramayana.

He was a tall and imposing-looking man and fond of playing practical jokes; he often frightened young Pot, creeping up on him unawares and giving him a scare until the boy grew old enough to anticipate his moves and laugh instead of cry. As he grew up and overcame his fear, Pot came to feel a deep affection for Khun Daeng, who, in turn, became fond of the young boy. He often called for him when Nai Pot returned from school in the afternoons. He liked to have him by his side when he tended in the orchids, or when he was in the house, or up in the studio, sometimes to run errands for him but mostly to watch and learn and listen. Often during the weekday there was no one around but Khun Daeng and the servants. His son was in Singapore expanding the family company, his grandson finishing his education in England. Khun Daeng found in Pot a lively companion and he encouraged the boy to study hard and to be a kind and considerate person.

Pot enjoyed being with the old man, who exuded a sense of fairness and decency, as well as generosity, even though his stories did ramble, and when he expounded his views on topics beyond Pot's early comprehension.

Khun Daeng did not seem to care if the boy was paying attention or not. Much of the time he seemed to be talking to himself, complaining about the way Thailand was going, bemoaning its loss of integrity and honesty and sense of justice and decency. 'I worked hard, Ai Pot. But I never cheated anyone and I tried to live by my principles,' he would repeat. Khun Daeng was a committed Buddhist and liked to try to teach Pot about the Dhamma, his monologues peppered with Pali terms the young boy hardly understood, though he listened anyway because he was entranced by the melodious, rasping voice droning on in the hot afternoons.

Sometimes, when they were in the studio, they would share a joke about a family member, or a guest who had been to the house that weekend, and Khun Daeng would get him to mimic the subject of their amusement, delighted by Pot's powers of observation and sensitivity to the nuances of body language and the quirkiness of speech patterns, and he would laugh until tears came to his eyes.

Nai Pot was aware that Khun Daeng wanted to help him to do well and prosper in the world, and he too wanted to make the old man proud of him. But even as a boy he was possessed of a contented nature, finding happiness where it was, rather than looking for where it might be. He had no ambitions. Khun Daeng respected that, just as the boy respected the old man's efforts to encourage his progress. Pot liked living in that big house, sharing a room behind the kitchen with his mother, listening to Khun Daeng's stories, jumping in the canal, and indulging his one true passion; cars. Khun Daeng had several, including an old Jaguar and a vintage Alfa Romeo, and there were others the family used as well as the company vans. Everything about them fascinated Nai Pot, who liked to help the drivers wash them down with soap and water and clean the interiors every day. HE would go with Khun Daeng to the garage to watch the mechanics repair or service them.

When Nai Pot was in his late teens Khun Daeng persuaded him to attend evening classes in English, which he enjoyed, at first. But the idiom and the grammar confused him and from listening to the foreign guests who came to the house for drinks or dinner on weekends his ears picked up not one English language but many: Americans and Canadians and Australians spoke English, and so did the British, but not the same English, and Indians and Singaporeans spoke English too, though Pot could barely understand them. The more he learned the less he understood and eventually, despite Khun Daeng's encouragement, he lost his motivation. Cars were what he wanted to be around and he determined that after his military service, and without telling anyone, he would try for a job as a driver with the Wongpanich firm Siam Land.

Being a member of the household he walked into his first job without any difficulty and began to ferry workers back and forth to various New Siam sites around the city. On weekends, he had to drive members of the family to the supermarket or the golf course or into the city to the cinema, or run errands for them as required.

Nai Pot enjoyed being at the wheel. It gave him a sense of control and freedom. He did not mind the traffic jams or the delays or the waiting around. He developed an instinct about other drivers and could spot the incompetent and reckless headed for a crash, and so give them wide berth.

His only regret was that the job took him away from Khun Daeng, who died suddenly after a brief bout of pneumonia. After the funeral, Nai Pot was asked to drive for the family full time, which he took to be an honour. And when Khun Daeng's grandson, Khun Sompop, left the business and entered the political scene, he asked Nai Pot to be his personal chauffer, which meant more work, longer hours and hardly better pay. He didn't mind because he was excited and proud. He had already heard the rumors that his boss was being tipped as a future prime minister. In keeping with Khun Sompop's new career there were more guests at the house now – company directors, celebrities, journalists and ambassadors – and after these gatherings Khun Sompop's wife would sometimes ask him to drive the guests back to their homes.

Driving for Khun Sompop taught him a great deal about Thai politics just by listening into the conversations he overheard being made in the backseat. He learned how alliances were forged out of an elaborate system of bargaining over power and interests, how acts of treachery were elaborately planned, how defenses were set up in advance of moves to offset the possibility of being found out. He realised the public knew only what the politicians wanted them to know and in the shadows existed a parallel and often opposite reality. He learned that goodness, dignity, generosity – values taught to him by Khun Daeng, Khun Sompop's own grandfather –were mostly absent in Thai politics, which was all about greed and ambition wrapped in fine rhetoric. When a politician talked about the good of the country, he was either deluded or lying, and Nai Pot would have been right to be cynical, but he held to an innocence sustained by Khun Daeng's pragmatic Buddhism: that all was Aniccam, which is hard to translate into English because it has so many meanings in Thai, but means something like 'impermanence', or 'transient'. In Thai politics it meant what goes around comes around and that everyone gets caught in the end.

During this period Nai Pot sometimes daydreamed of being rich enough to own a really fancy car of his own. He had once seen a Ferrari drive up to the house and for months after that he had collected photos of the different models in the auto magazines and visited a motor show where they were featured. The beauty and elegance that he saw in them made his mouth water. But at the same time he accepted that unless he won the lottery, or robbed a bank, it would be a treasure always out of his reach. This did not bother him. He had learned of people lending money from the loan sharks just so they could buy a large TV that hardly fit in their room, and the subsequent trouble they'd be in when they could not pay their debts. It seemed so pointless. He was content to dream. And besides, as time went by he came to understand that none of the people were any happier for it, including his boss Khun Sompop who could afford (and did own) so many shiny, expensive objects and amazing cars. Their stressful lives and their constant preoccupation with money and power did not make him envious.

He drove for Khun Sompop for five years and witnessed his boss climb to the top of the pile. When Khun Sompop left politics, Nai Pot was no better informed about his motives than the next person, so surprising was the move. They had driven out of Bangkok for a meeting at Sia Oui's estate in Khao Yai and the next morning, without explanation, Khun Sompop told Nai Pot he was quitting politics for good, that he was leaving the country for a while, and that he and the family would no longer need his services. Nai Pot was hurt, mostly because his dismissal, after his years of service, was so abrupt and came without proper explanation. But he did not protest or complain. Intuition told him that Khun Sompop did not want any reminders of whatever had happened. At this point Nai Pot was in his mid thirties with experience and a clean slate – no accidents, not even a parking ticket to his name. Getting another job was easy. It was time to move out and find his own direction, especially since Khun Daeng had gone and when Mae Lien, his mother, had decided to return to their village in her old age.

In fact, he later discovered that Khun Sompop, either from kindness or a guilty conscience, had fixed him up with another job. But before taking up his new employment Nai Pot accompanied his mother back to Isaan on the train. It was a chance to visit the village where he was born and to attend a wedding while he was there. Having only once before made the journey as a young boy, when a relative had died, he hardly remembered what the place or the surrounding countryside looked like. It was a long, uncomfortable journey and he had been nervous of going. Raised in Bangkok, the city was familiar to him; the noise, the colours, the convenience of everything. In comparison, northeastern Isaan from a distance seemed primitive, deprived of the lifestyle on offer in the capital. Yet when they reached their destination, quite unexpectedly he found that he was touched by the beauty of the open landscape that contrasted so vividly with the hardness and contamination of the Bangkok streets. The rice was ready for harvesting, the trees were full of fruit. There were fish in the river. The air smelled fresh and clean. His mother had always talked to him about Isaan with such passionate loyalty and now he discovered that what she had told him was true; the warmth of the northeast and its people, his people, the slowness of the lifestyle and its simplicity was enormously appealing. He was treated like a member of the community whose presence had been missed.

In the weeks following that brief trip Nai Pot could do nothing but dream of returning there. Isaan had opened his eyes, and when he was back in Bangkok he no longer saw it as the place he belonged. He reflected that the city he'd known had all but vanished, and in such a short time. The canals were now black with sewage and the boats no longer went by along them. The busy, colourful markets he used to go to as a child with his mother were no longer there. Instead everyone shopped in bright neon-lit megastores; more efficient, fixed prices, no haggling, but ice cold. Beautiful old buildings had been replaced by ugly, impersonal, high rises. He heard that even the old Wongpanich mansion was going to be sold and turned into a luxury spa complex. And it wasn't just the way Bangkok had been physically transformed. After the warmth of the villagers he realised that what was missing from the city was kindness. In their rush to fulfil their dreams people were forgetting how to treat each other with basic human decency. Common courtesy and politeness and sincerity were fast disappearing from everyday exchange. He witnessed this with his own eyes and had several times had to deflect the rage of impatient drivers.

His mind kept playing with the possibility of trying to make a life in Isaan. It seemed to him this would complete a cycle that had begun with his mother's journey to Bangkok. But what would he do if he went back there? He had no skills as a farmer. They had joked that his hands were soft as a city dweller's, and they were right. In the rice fields his soft body would crack like parched earth under the blazing sun. He was in the same position as most of the other people from Isaan, who were stuck in Bangkok, not wanting to be there but unable to return to their birthplace, with nothing but the nostalgic verses of their loog toong music to sustain them.

IV
Taew Finds Love

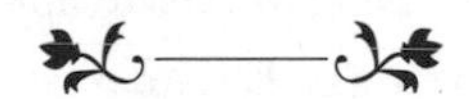

Taew sits in her wheelchair nursing a glass of white wine in front of the evening news. Heavy rains expected in the north, possibly producing flooding, more trouble in the south; a temple on fire, soldiers in full combat gear flagging down pickup trucks. Violent images, but they barely register. Then there is a piece of breaking news; the suicide of a well-known writer. Policemen are standing outside an apartment block in Thonburi, and paramedics are carrying a body on a stretcher along the corridor of a hospital. The face of the dead man is not completely covered, revealing a high forehead and a mess of greying black hair wetted with blood.

Taew switches the TV off with the control button. She can do nothing about what she sees on screen. Nor can she identify, as she might have done once, with the suffering of others. She is too wrapped up in her own day-to-day struggle with pain. But the writer's suicide disturbs her.

Nai Pot stands under the shower and washes the bitter grime of the city from his skin and hair. *Lobha*

-- *greed – where does it come from?* This is the question on his mind. *Why are some people greedy and others not? It doesn't seem to depend on how much or how little they have.* He wishes he could ask Khun Daeng, as he had done as a child. He could hear the raspy voice in his ear: *Lobha nak, mak Larb hai. With greed, good fortune disappears.*

Earlier that day he had driven a young Russian couple to various gem factories around the city where they had insisted he bargain for them, and then through heavy traffic to lunch in a Japanese restaurant recommended by their guidebook, and then to a shopping mall before returning to the hotel. The traffic had been especially bad due to heavy rain, the places new and badly signposted. At the end of the long day's work they did not even bother to thank him, or tip him.

The other hotel drivers had often told him how to treat the farang to ensure they at least buy you a meal or offer you a beer during the outing, and tip you well besides, but after two years Nai Pot figures it's a trick he'll never learn: how to tell the stingy from the generous.

The old man would have enjoyed discussing these issues with him and together they would have reached some conclusion. But alone Nai Pot feels frustrated. He is too tired to figure it out and he is late for his meeting.

Taew cannot avoid the self-pity that sweeps over her. The evenings are difficult, especially Friday evenings. Normally she would have been preparing to go out to dinner in a fine restaurant, and then to a bar or nightclub. She misses the music and the dancing and the company of the other women like her who know how to enjoy themselves, and the admiring looks of the men as they wonder what they had to do to touch her silky skin and possess her lithe body. That life is over. It will never come back.

But why was she punished so cruelly for it? Had she been so bad? No. She had merely learned to survive.

At sixteen she was seduced by a handsome, married policeman in her village who treated her no better than a prostitute. She knew all he'd wanted was to fuck her, but she was infatuated enough to enjoy their sleazy adventure – for a while. When his wife discovered what was going on she threatened to expose him and kill them both. He gave money to Taew and told her to pack her bags and leave the village. No one was sorry to see her go except her father, a Chinese tailor, who was losing a worker he didn't have to pay. He had remarried quickly after her mother had died and there followed three children in quick succession. Taew was just another mouth to feed.

She headed for Bangkok aware that with her pale skin and attractive looks she could have found an easy-money job in the entertainment business. She knew all about it from the girls who'd returned to the village. 'You're spoilt for choice if you're pretty,' they'd told her. 'All you need is a doctor's certificate to say that you haven't got Aids, and you're in.' But even though she was tempted Taew was ambitious and wanted something more challenging than selling her flesh. Anyone could do that. She wanted to explore other options and that meant learning about what made Bangkok tick. She began her first job there as a waitress in a restaurant and found a room with one of the other serving girls. On her free days, she attended a secretarial course. Taew often joked that work was in her Hakka blood; she had been sewing clothes in the back room for her father from the time she was tall enough to use the machine. She did not mind hard work. She wanted freedom and liked the tempo of 1980s Bangkok, which pulsated to the rhythm of twenty-four-hour days as it was slowly transformed from sleepy Asian backwater into a modern metropolis.

Her looks and glowing references from the course instructors landed her a job as a secretary at a Japanese carmaker. Within a few months she was having an affair with the general manager who offered to marry her and take her back to Osaka. She changed jobs and lovers and learned the ways of the city, and of men. Both required skill.

The pace of development in Bangkok was relentless and she learned to avoid getting stuck in traffic jams as more and more cars poured onto the inadequate roads, and to avoid being stuck in disagreeable situations with men whose prospects were inadequate, or whose demands were impossible to meet. She wanted nothing from them but pleasurable sex and pleasant company and if they wanted to offer a generous gift or thick envelopes of cash – to have fun – then fine, so long as they let her be free. But soon they felt threatened by her freedom or felt the need to possess and control her; then would begin the whining and the moaning. Before long, the man would be talking of chucking his family or threatening to do something equally foolish to have her. This she found both annoying and unnecessary.

Taew knew what she wanted, and after a number of unpleasant experiences she came up with an escape plan, a pre-emptive strike that left the way open for her to leave a relationship before it reached crisis point.

It also made her partners work all the harder, thinking she was married to a senior policeman and was a mother with two children. She used the name of the cop who took her virginity, and the names of his children, and enjoyed the role she played, which let her share the frisson of risk her men seemed so much to enjoy. She liked keeping things careful and clandestine, so she controlled the when and the where; and when time came to end the affair there was never any argument. Her children came first, and her policeman husband, rising higher and higher in the ranks, was the jealous type, not one to be crossed.

Over the years few of her lovers had stood the test. Occasionally one would come along who held her interest for a while because he was more intelligent and witty and could make her laugh. But sooner or later she would use the married-to-a-cop excuse to find a way of ending the affair. Often, with these more intellectual types the separation was painful and acrimonious. One of them sent her a farewell message on her mobile phone saying:

'You are a bad woman. I don't know why I've wasted my time suffering over you. You are a liar through and through. You have lied to me, but you have also lied to yourself. Can you live with that? You are as cheap and immoral as the rest of them. I thought you were different.'

Her fortieth birthday crept up on her like a tiger in the night, from behind, unnoticed until it was suddenly on her. She began to notice that she no longer turned men's heads like before. They wanted young, firmer flesh. Before long she would be a good-time girl doing the chasing rather being the catch. And then, like a gift from the deva realm, Khun Prayat, the millionaire businessman arrived – a chance at a secure and comfortable future. Goodbye short-time hotel rooms and borrowed apartments. Hello married life.

Nai Pot drives his pickup with extra care that evening. The rain pours down and he sees a motorbike skid into the side of a truck. The rider, no helmet, doesn't get up, and the truck doesn't stop. People are rushing over. He watches the steady swishing of the powerful windscreen wipers, the radio news on low, but he is not listening. From experience he knows there is nothing to do but wait, and yet tonight he cannot keep his rising impatience in check. He cannot get the stinginess of the Russian passengers out of his mind. But he knows that they are not the true cause of his discontentment. Ever since his trip to Isaan he has lost the enthusiasm for his job. He no longer enjoys exchanging gossip with the other drivers, or standing around in some carpark in his starched uniform waiting for his passengers to finish their meal. All he can think of is how he can manage to leave the city and find a way of setting up a small business and not be at anybody's beck and call.

Nai Pot accepted his new job gratefully but he did not feel the same respect towards his new boss, who was a self made man without the culture and breeding of Khun Sompop. At first he found it difficult to adjust to Khun Prayat's blunt, aggressive, no-nonsense manner, but as the months went by he became used to it. After all, he was only the driver, and the generous pay and overtime bonuses Khun Prayat gave him more than compensated for the rudeness he had to endure from time to time.

He had been driving for Khun Prayat for eighteen months when he was directed to pick up 'the woman I am going to marry'. Nai Pot covered his surprise skillfully and offered his congratulations. He concluded that they must have met at some party to which Khun Prayat had driven his sports car, a pastime that had become more frequent in recent months. Khun Prayat added that Nai Pot would now be her driver, and take her wherever she needed to go. He would, during working hours, use one of the party's cars.

Nai Pot nodded his assent, which did not mean he was happy about the new arrangement. In fact, on being told the news he was, if the boss had bothered to ask, distinctly unhappy. Nai Pot was not superstitious and did not believe in premonitions, but the night before it had come to him in a dream that something evil was going to happen. In the dream he was floating along a wide, green river. There was a golden light. A beautiful young woman flew down from the sky and began to swim next to him and then to seduce him, but a Naga appeared and reared its seven heads and took the woman away.

When he saw Khun Prayat's future bride in the flesh, he knew immediately that the message contained in his dream was going to be made manifest. His guard was up from the moment they met. He was suspicious of this ambitious social climber and of her blatant efforts to please and satisfy his boss. He tried not to notice the way she would grope Khun Prayat on the back seat. Yes, they were going to be man and wife, but it seemed improper for such a public display. Nai Pot also tried to ignore the fact that as soon as he saw her he was consumed by lust.

Nai Pot had little knowledge of sex. His first experience had been at the age of seventeen, when the daughter of the Wongpanich head gardener had taken it upon herself to climb into bed with him. She was several years older, and was high the first time, but not so on her subsequent nocturnal visits. He kept expecting her father to storm in and deliver a thrashing, but no such thing happened and after about a year she disappeared. He never saw her again. In the army, he went along with the other conscripts to a brothel near their barracks but, apart from the physical release, he derived neither pleasure nor satisfaction from such encounters. It wasn't that he'd felt himself superior to his comrades, some of whom were boastful and tried to project a macho image; some of whom, like him, were awkward about the whole thing. He simply thought that having sex with someone you didn't love and didn't know, who probably didn't want to be there anyway, who'd opened her legs thousands of times before to others and would continue to do so thousands of times after, was not something one should take pleasure from. After the first few visits he made excuses and begged off joining his companions. Years went by and he never found anyone he liked enough and gradually, like a monk, he stopped thinking about it all together, until the day he met Taew.

She was unashamedly sensual and he could not stop himself from noticing every single detail about her appearance; the short skirts, the long straight hair reaching to the small of her back, the pale skin, painted nails on fingers and toes. He fantasized about her, how it would be to make love to such a voluptuous woman, and he imagined that she was perfectly aware of his feelings. Not that she ever flirted with him, or led him to think he could even dare to overstep his role. On the contrary, she was always polite and correct, keeping her distance even to the point of using the same honorific that he used to address her, calling him 'Khun Pot' rather than the more familiar 'Nai Pot'. But the more respect she showed, the more intense became his feelings.

He was in such turmoil that he brought forward plans to set up his own business of showing the more curious tourists what lay beyond Bangkok. He knew other drivers had made similar moves and that there was already a crude infrastructure allowing for tours of several days in comparative comfort. But he was still not quite ready. Another couple of years of saving was all he needed.

Even so, when he was on the verge of resigning, Khun Prayat took Khun Taew on a holiday to the US, where they were to be married in a quiet ceremony. Nai Pot retreated to a temple in Ayudhya. With practice and chants and gruelling sessions in the herbal sauna and the guidance of a monk, who was totally aware of what he was going through and showed both sympathy and compassion, Nai Pot saw that he had been temporarily under Mara's spell. His fiery lust was calmed and he came to accept that Khun Taew, like the Ferrari he once coveted, would always be inaccessible. By the time the couple returned he had stopped seeing Khun Taew as an object of desire, and began to see her as a person. He found himself appreciating her astuteness and her sense of humour. She continued to treat him with a natural friendliness that gradually developed into trust. He was now her loyal driver. Balance had been restored.

Over the following year, Nai Pot grew concerned that Khun Taew was becoming depressed and this he put down to Khun Prayat's desire to have children. A visit to a private clinic revealed she would never be able to do so and, with that simple truth, her world changed. She grew bored with shopping and going to spas and dining out in fancy hotels and restaurants with her husband while he began to fill his schedule with conferences and parliamentary committees and golf. From other drivers Nai Pot learned that Khun Prayat had a mistress and was grateful he was not drawn into collusion. When together in the back of the car he could hear them having little arguments, although they kept their voices low. Khun Taew seemed agitated when she was alone and was often on her phone. He overheard her once tell someone – a friend from before she was married – how much she wished she was working again, that she was fed up with her job as a rich housewife, that all she did now was decorate her husband's arm and attend grand functions. She resented that he never sought her counsel, nor permitted her to speak her mind in public, save to compliment him on some comment or observation and laugh at his jokes.

Nai Pot did not understand. She lived a life most could only dream of, and had found nothing in it that made her happy. He hoped, for her sake, that Khun Prayat would grow tired of her and set her free to find her way in the world.

Khun Taew must have known her husband was already straying and had determined it was a game that two could play, because a light returned to her eyes, notably after her visits to a newly opened private sports club became a thrice weekly routine. On these afternoons Nai Pot could expect two hours to himself, which he found useful for running personal errands or talking with other drivers, especially those who knew the countryside and interesting places that might appeal to his future clients.

One day he took a call from Khun Prayat, who wanted to talk to Khun Taew, but her phone was off. Would Nai Pot have her call him back immediately? He was agitated.

Nai Pot immediately went to the club where he had dropped Khun Taew off an hour earlier. He couldn't find her and, thinking she might have left without telling him, he went to the reception desk to see if she had left him a message that she had gone off to the cinema or to a shopping mall with a friend, as she sometimes did, in which case he would be free for the rest of the day.

There was no one at reception and, as he waited, he heard Khun Taew's voice from beyond a STAFF ONLY door, through which she came tumbling in the arms of a young man dressed in shorts and a polo shirt. Nai Pot hesitated, embarrassed for having seen something he should not, and uncomfortable at the embarrassment he had caused. He fled to the car.

Khun Taew said nothing as she waited for him to open the rear door, and did not thank him for the message to call her husband, with whom she conversed in a light bantering manner. She said nothing during the drive home.

She was tense the following day, and Nai Pot feared the silence might go on forever when, as he drove her to a hotel near the river to pick up Khun Prayat, she leaned forward and touched his shoulder gently and said: 'Khun Pot, I am grateful to you.'

He did not want to become her accomplice and he would not lie for her. What she did was her own responsibility. But she seemed alive and radiant again, and he preferred this Khun Taew to the unhappy one.

The shooting came three months later.

On that particular afternoon, Khun Taew had told Nai Pot to fetch her not from the club, but from the Oriental Hotel later that evening. He knew this meant she would be spending the afternoon with her lover. Khun Prayat was in Korat, campaigning in support of a local politician in the upcoming elections and would not be back until late. Nai Pot was grateful for the time off.

After dropping off Khun Taew he drove towards the temple in Onnut, where his cousin was visiting her daughter. He had not seen Mae Jom since he had visited the village. She was in Bangkok to fetch her daughter, Tookta, who had AIDS. Nai Pot had the temple in sight when Khun Prayat rang. His voice was slurred, and Nai Pot could hear the clink of ice cubes against crystal.

'Where's my wife?' he demanded.

Given that was Nai Pot's job to ferry her all over Bangkok, he couldn't say he did not know, nor could he betray her confidence.

'She is at the sports club, Sir,' he said, regretting immediately his lie.

There was a long silence. Nai Pot heard the clink of bottle against glass. 'Leave her there, then,' slurred Khun Prayat. 'I'll be back late.'

Nai Pot was puzzled by the point of the call, thought they were perhaps having another of their simmering disputes, and gave it no further thought.

He was looking forward to spending some time in the temple and thought he might take the opportunity for some meditation of his own. He was ignorant of subsequent events.

Khun Prayat knew everything about his wife's infidelity because he had hired private detectives to follow her. He had been irritated by her unhappiness and could not tolerate her sullen moods. But her newfound happiness had made him suspicious until it gnawed at his marrow and sucked the pleasure from his life; he could no longer enjoy the delights of his mistress, who exacted a high financial price. His detectives reported to him on a daily basis, and that was how he learned name, time and place. A text message gave him an address and, having slipped back to the city that morning, he made his way to the hotel.

As the shirtless man opened the door, Khun Prayat shot him three times in the chest, the hot shell casings pinging on the marble floor as they were ejected. He stepped around the man's body and faced a frightened Khun Taew, who clutched a light cotton bed sheet uselessly to her chest. He fired twice, once in each of her knees, and put a third bullet in her belly.

The sound of gunfire is not uncommon in Bangkok, but the luxury Oriental Hotel promised protection from such things. The sound of running feet preceded the arrival at the door of security guards, guns drawn, and behind them a crowd of guests whose curiosity overcame their fright. Reporters gathered quickly from nowhere, like carrion at the security doors, and as police led Khun

Prayat from the building, they caught the words: 'No one takes what is mine. No one takes what is mine.'

Nai Pot was interviewed four times by various departments and each time he said, 'I am only a driver.' Under pressure, he added, 'I am not told anything except where to go. My employers' private lives are their business. It has never interested me.'

When an investigative reporter offered him a substantial amount of money for a tell-all story, Nai Pot said, 'I know nothing. I was told nothing, I'm only a driver.' Finally, they left him alone.

Nai Pot visited Khun Taew daily in the hospital, at first in the company of other household staff, some of them praying as they looked through the glass window of the intensive care unit at her body riddled with tubes, an oxygen mask over her beautiful face. None of the staff liked or respected Khun Taew, but it would appear unseemly if they did not participate in the vigil, and it satisfied their morbid thirst for gossip. Some of them had known Khun Prayat since he was a boy and did not hide their feelings of blame and disgust at what Khun Taew had done to him.

After she was out of danger and moved to a private room, Nai Pot was her only visitor. When she had recovered enough to speak, Khun Taew thanked him for being so kind. In the following week, she was visited by a series of lawyers and Nai Pot learned that Khun Prayat's family had cut a deal. So long as she did not interfere with the company and left the family home, she would receive a generous pension to compensate for her suffering. They would agree not to pursue the matter further.

'They're treating me as the criminal,' she told him after one meeting. And "compensation"?' She was crying. 'The doctors say I will be in a wheelchair for the rest of my life, and that my insides were torn up so badly I could still easily die.'

Nai Pot held her hand and said nothing.

After Khun Prayat was tried, convicted and sentenced to life imprisonment his family was furious. Their expensive lawyers and bribes and arm-twisting got them nothing, and they took it out on Khun Taew. Instead of the agreed settlement, she received a small apartment, a live-in carer, and a monthly allowance to cover her food expenses. When she complained, Khun Prayat's younger brother spat on the ground and said she should be grateful because, if it were up to him, he would have her thrown off the balcony for what she had done to his brother.

'Is this what I deserved, Khun Pot? Was I so bad?'

He had no answer. He had gone to the temple and lit incense and talked to the same monk who had guided him through his earlier turmoil. He told him that he was trying not to hate, knowing it to be an emotion that consumes too much energy without giving anything in return. But despite his best efforts, he had grown to hate the city, which he believed to be a monster that devours everything good in a person, leaving them broken and empty and dead.

Taew wants another glass of wine and calls out to Fo, but there is no response from behind the closed door and she knows the girl is either on the phone to her boyfriend or pretending not to hear. Fo no longer bothers to hide that she is fed up with the job and now often tells Taew she is not going to waste her life looking after an invalid. Fo is paid by the family, so there is nothing she can do.

She keeps thinking about what she has just seen on the TV, the writer's suicide. The option to finish with it all has been with her since leaving hospital. Before then she could never understand anyone would ever consider ending their own life. But now that she was bound to a wheelchair, her beauty gone, her sensuality a dusty well, she begins to understand. And though she knows it is useless, she often wishes Prayat had aimed higher and finished her off properly.

But she is not ready to end her life, not yet. However hard the journey has turned out to be, the urge to see what will happen ahead, the curiosity of what is around the corner, still keeps her going. And, besides, she does not want to hurt Pot, who still addresses her as Khun despite the fact that she has asked him to address her more familiarly.

She is glad he is coming to see her. Tonight she wants be honest with him, and tell him that she has never felt the same about any other man. It is time he heard her express these things outright without any elaboration, and he can respond to her how he chose. She expected nothing. They were a world apart. He was younger than her, and moderately attractive in a quiet way. He could easily find himself a nice girl. But she held onto hope. She could still do that.

She wheels herself in front of the long mirror that hangs on the wall by the window, put there by some previous occupant to correct an imbalance in the room's feng shui. She examines her reflection, patting her hair as she does so, wonders if she should put on a little lipstick. She shakes her head at her vanity and smiles.

Outside, the rain falls relentlessly, as it has been doing for weeks. The wet road below reflects the bright lights of the neon shop signs and hoardings; headlights blur through the sheets of water cutting down from the sky. One day the city is bound to be under water.

She has had enough of Bangkok. She has devoted her life to its pleasure and now she wants nothing more than to leave this heap of concrete and refuse and go somewhere quiet where she might find something left of herself.

Pot is a kind person. That much was always obvious. But his constancy and loyalty has surprised her, and lately she feels their relationship has entered another phase. But is there a life in it? And does she even know what life is? She had wanted everything on offer in Bangkok, populated by its dreamers sleepwalking through their routines, willing to sacrifice integrity and intelligence for the sake of those dreams, elusive and intangible. She refused to join them and her affairs were an attempt to take control, to pretend that she could choose. It was an illusion nevertheless.

When she first arrived in the city she briefly toyed with the idea of training to be a teacher, or a social worker, of making some contribution. But by the time she'd met Prayat those ideas had long ago faded away. Lately her altered circumstances have her thinking again of the possibility of doing something meaningful. But if she were to take this path, she would not be able to do so alone, not in her condition. She would need a friend like Pot, someone she could depend on and respect. If he were brave enough to care for her it would keep her to her task, and give her a reason to go on.

Tonight Nai Pot's mind is made up.

Khun Prayat' cruel act, and what his family subsequently did to Khun Taew, convinces him that he has to take her away from Bangkok, while there is still time, and return together with her to the northeast where there is still space and air and green growth. He sees a future for them there, humble though it might be. He has enough money saved to buy a decent van for his small travel company. He is willing to work hard but he does not want to do it without her. He will not leave her alone in Bangkok. There is no great barrier to cross, no walls to break down. She knows that he cares for her and she, so wise to the ways of men, must know how he feels about her. All he needs to do is tell the truth and reveal what is in his heart.

It is as simple as walking through a curtain of rain.

Part Three

Bangkok

Chapter Six

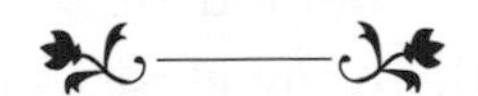

Clare scans the skyline framed by her hotel room window. She is fresh after the shower, a small bottle of water in one hand, a rose apple in the other, white bathrobe knotted loosely at her waist. She looks down at the sprawl of the city and fails to recognize any of it. Her eyes dwell on the bruise of rain cloud gathering over the electric green vegetation in the distance. She recalls how she'd once welcomed the monsoon skies and their promise of relief from the intense, damp heat of the city. She especially remembers how clean the streets felt after the rain, and the smell of the trees and flowers. Now, according to the news, there was nothing benevolent about this weather. They were waiting for the flooding to begin.

She sips her water and bites into the apple she has washed and polished from the fruit basket, and delights in the crackle as her teeth cut into its crisp, pearly white flesh. Her palate tingles. It has been years since she has eaten one. In London she never ate apples because she found their waxy, chemically enhanced appearance unappealing. This rose apple's blood-red skin is firm, she notes, and its taste is altogether unfamiliar. Its

Thai name rises from the depths of her memory: Chompoo.

At the hotel reception they addressed her as Mrs Stone. She had not bothered to correct them for fear of upsetting the check-in procedure with a relatively minor matter; she'd wanted only to reach her room without delay. As the door closed behind the boy who had brought her bags and pointed out the room's various switches and controls, she let herself relax. Dizziness swept over her. It was to do with the thickness of the carpet, which suddenly seemed to turn liquid. She felt her weight sinking into it and began to lose her bearings altogether. Her body started to shake slightly. Only an echo of a command played in her head. She blinked her eyes fast before closing them and conjured up in her mind the image of a thin red thread, which she now held in her imagination between the index finger and thumb of her left hand and gently pulled it in. Airport. Plane. Bossy attendants. Clouds like dark mountains. Movie on minuscule screen, romantic comedy, title forgotten, not important. Immigration. Uniformed man holding up sign with her name. Clare Stone.

Reconnecting with her name, she felt a rush of physical relief, like breaking the surface after diving deep into water, lungs bursting for air. Everything flooded back and the world was normal again: no trembling, all solid. The television was in the corner, the mini bar next to it, framed picture of old Bangkok on the wall. She told herself she was overtired from the journey and confused by her new surroundings, which were not as she had expected. The mind game still worked, proof enough she had a few more miles left before she'd need to rely on the kindness of strangers, her life the responsibility of others.

She had decided not to tell the consultant that she was leaving for Bangkok. If he had known that she was going to travel alone six-thousand miles to a tropical city, in a country that was about to be flooded, he would probably have said something unpleasant to try to discourage her. But she had made sure, just in case anything did happen, that Joanna knew where she would be staying. A day before the flight she'd had a moment of doubt.

She'd gone to Paul's office and told him: 'Perhaps it's better I don't go. After all I don't need to. I can email him...'

Paul looked at her aghast, as if she had uttered some heresy: 'Business class to Bangkok? The Oriental Hotel for five days? Are you mad, Clare? Just enjoy it!'

His shoulders were shaking with laughter as he walked away.

He had finally pressured her to tell him exactly why it was important, as she claimed, for her to meet Tarrin and question him face to face. And so she had told him that when she had read Tarrin's first story about Mae Jom she had come across a passage that struck her as more than a piece of coincidence. It was when Tarrin described the scene in the restaurant when the waitress witnesses a near-fight between a group of foreign diners and a bunch of American GIs. Clare had no doubt in her mind that she was, in fact, the foreign woman. The description of the place, and the details he included – the people around the table that night, the clothes they wore, even the food they'd ordered, were uncannily precise. Then there was the incident with the gun. It was impossible that he could have imagined it. Had he been there? Or had he obtained the information from someone else – the waitress perhaps, or one of the foreigners? For Clare it was so odd that Tarrin, writing fiction, had captured so vividly this particular moment of her past, and in doing so reminded her of the young woman she had once been. She felt she had to meet him in person. Besides, she added, an interview with the author would help promote the book when it was published. Paul had accepted her arguments and persuaded the directors to pay for her trip.

What she didn't tell him was that Tarrin was a subplot, not the real reason she wanted to return to Bangkok. It was true that she had been surprised to come across the passage in his story. But it wasn't just the sense of serendipity it produced. The incident meant far more to her than that. It was the memory of Mickey, and with it the disturbing emotions she had fought for years to keep at bay. It was because of Mickey that she had gone to Bangkok the first time, and it was because of him that she now needed to return.

Mickey, you bastard! Clare tosses the apple core into the bin under the desk and turns to the mini bar and its enticing stock of beverages and snacks. A gin would be nice, she decides, and finds a little green bottle of Tanqueray and a yellow can of Schweppes. The fridge clicks twice and the thin paper price list floats down to the carpet to remind her of its outrageous prices. A gin and tonic now and a dry martini before dinner in one of the hotel restaurants, with perhaps a glass or two of wine, and she would be ready to let sleep reset her circadian rhythm to this new time zone. She would wait until the morning before trying to contact Wandee.

She mixes her drink, and toasts its medicinal properties. 'To juniper berries,' she declares out loud, and takes a long pull on her tall glass, drinking off a good third in the first hit. The doctor said mixing alcohol with her medicine was not advisable, but soon enough she wouldn't even know the word gin, let alone recall why it was important.

The next morning Clare gets up and begins to apply her make-up. At the magnifying mirror she gently rubs cream with the tip of her fingers into the wrinkles around her eyes. She bares her teeth and checks her gums. She blow-dries her now uniformly grey hair. When she'd stopped colouring it, a streak appeared that her friends claimed to be characterful but which she found annoying because it made people stare.

She cherishes this morning ritual, a quiet time to reflect on who she is: at a late stage in his Alzheimer's her father could not recognize himself and several times she had found him standing in front of a mirror, petrified. In a hotel lobby, he had jumped behind a faux marble pillar and whispered: 'It's him ... the one who's after me.'

The pretty girl at the desk in the business centre tells Clare for the second time there is no answer at the number she'd wanted and asks, again with a polite smile, if it is correct. Clare looks down at the number Paul had given her. Secretly she is hoping that it is wrong, or has been changed, and that she can go home. But on the following try, the girl beams a smile and gives a silly thumbs-up, but, as the call progresses, she looks flustered. She puts her hand over the mouthpiece and says: 'He's asking what your name is, and how old you are.'

Clare gives her eyes an upward roll. 'My name is Clare Stone. He knows I'm here to see him. I sent him an email.'

The girl looks even more confused, then she giggles, covers her mouth, covers the mouthpiece again. 'He says he's not well.'

'Tell him neither am I,' said Clare unable to hide her impatience.

After the next exchange the girl relays: 'He says you'll have to go to his place as he does not leave his flat anymore.'

'Where is it?'

'It is on the other side of the river, to the west. It's about half an hour away by taxi, depending on the traffic.'

'Thank you. Tell him I will be there in an hour or so. Make sure you write down precise directions.'

At the concierge desk there is some debate over Clare's destination and her request for a taxi. Several of the boys appear surprised, puzzled, that she, a foreigner, could have any business in that confusing, unattractive part of the city. Clare finds it all unnecessarily complicated. *If it's so hard to find the right address, how on earth will she get back?* They pick up on her nervousness.

'Perhaps Mrs Stone would like to make use of one of our drivers. He will wait and bring Mrs Stone back to the hotel when she has concluded her business?'

Clare nods. 'Very good, but we must go now.'

The driver is formally dressed in a pressed, white-cotton suit with matching hat. He smiles as he opens the door for her. Clare tries to get a look at his nametag but she can't make it out. For a moment she imagines that it is Nai Pot. She thinks, Mickey would have made a joke of this: Mr Pot driving Mrs Stoned.

The ride through Bangkok starts slowly. That morning the roads are choked with traffic. The sky is metallic grey. Clare resigns herself to being late and tries to relax. She focuses on her meeting with the author. Her younger self might have been curious to meet a writer whom she had edited and whose work she liked. Sometimes in the past she had been disappointed; the gap between the person and the writer could turn out to be vast. With Tarrin she is neither excited nor expectant. She will reserve her judgment. All she knows is that he will be the last writer she will ever meet.

Tarrin Wandee, mid forties, single. Born in northeast Thailand, Mahasarakham, his family moved to Bangkok when he was three years old... She is familiar with the basics of his CV. A recent photograph of him showed a man possibly of Indian descent with large, piercing eyes staring straight at the camera and frowning with a serious, intense expression. He's self-obsessed, she'd thought, like most writers. But she enjoyed his writing, and, though she cringes at the cliché, she felt he had achieved one of those 'East/West' reading experiences. Wandee was both an insider and an observer, and he seemed at home in the different strata of Thai society. He had captured the sense of corruption that pervaded the city – the lives compromised by dark desires. But he had also managed to provide some ray of hope. This was what intrigued her.

As they cross the Chao Phraya River she strains her neck to see the boats below. She suddenly remembers taking a ride in a longtail, carving through the water at breakneck speed. Mickey is taking her to visit a temple and for a meal in a small noodle bar on the quayside. She thinks of the distance that separates the woman she is now, riding in the hotel limousine, with the shambolic young woman she was then.

When the driver turns off the main road past the bridge it is as if they have passed through an invisible timeline. Around them are small, squat dwellings and children walking naked and barefoot in the road, brown shouldered women in sarongs and flip-flops talking on street corners, old men smoking and watching, and an ice-cream seller pedaling her tricycle past clumps of trees. Everybody stares at the limo.

'Don't worry,' he says. 'They are curious. Why does a limo come here? You will be much talked about tonight.' His laugh puts her at ease. He stops the car at a coffee stall, lowers his window and asks the old woman there for directions. She takes her time to answer, intent on the coffee she is brewing, and waves a bone-thin, wrinkled hand in the direction the car is facing and rattles off a string of words.

'It's not far,' he says to Clare as he pays for a coffee he doesn't want.

'We are not far from the river. Look.' He points to the right, to a space between two buildings, and Clare can see the funnel of a huge ship.

'So different from the rest of the city,' she says.

'No one comes here. It's not beautiful.'

'But it's a proper village.'

'Not at night.'

'What happens at night?' she asks.

He doesn't answer, intent now on manoeuvring the long car past large, overturned plastic containers partially blocking the road. An old man in rags, picking through the heap of garbage, looks up and shouts something. His expression is angry and he looks as though he is about to attack the car before he then bursts out laughing, and bowing low to the ground in mock reverence. The driver takes them slowly down narrow, twisting streets and Clare awaits the sound of metal scraping against wall, but he is skillful. They are now escorted by little faces trying to peer inside to see the rich visitors. The next moment he gently brakes in front of a rundown apartment block.

'We are here,' the driver says, pointing at one of the blocks. 'This building. Third floor. I'll go up with you.'

'No. It's all right. I think I'll be OK. But you will wait for me, yes?'

He nods. 'Yes.'

Clare takes a minute to look at the building, a common grey concrete block constructed with cost rather than utility in mind. Her legs are tired and she is soaked in sweat by the time she reaches the door of number fourteen. The door is ajar, open enough for her to see inside – a man sitting on the edge of a large, untidy bed with his head in his hands, fingers buried in long, curling dark grey hair. He is dressed in nothing apart from his shorts, and there is an intricate tattoo on his left shoulder. He is still as a sculpture.

Clare knocks, but the man doesn't move. She pushes the door open, clearing her throat by way of warning as she does so, but still he doesn't move. She coughs again and waits. The room is the size of her large master bedroom in London. The floor is littered with yellowing piles of books and magazines in Thai and English. There is a writing table heaving with the curling pages of notebooks, a laptop, a desk computer and other technological paraphernalia. There are no photographs. Old Thai film posters decorate one wall. In the far corner is a tall, slim refrigerator, on top of which are maybe half a dozen unopened bottles of Mekong. Next to this is a small gas cooker, and on the other side a sink filled precariously with unwashed crockery. On a low table by the bed is a mobile phone, a packet of cigarettes, a lighter, a tea saucer overflowing with butts and ash, empty tins with peel-back tops, a teaspoon, some silver foil and a bottle of mineral water. The air is musty and faintly foul.

Clare shuffles in, and he acknowledges her with a deep, unhealthy sounding croak: 'I can't get up ... water ... from the fridge, please ... and bring me the pills by the sink.' The accent hints at Australia, where he had lived for four years when he was younger. Clare brings him a glass of water and the pills. His expression is one of defiance mingled with vulnerability. He looks up and thanks her.

His face is unshaven, its features smudged with illness or exhaustion. There are dark circles under his eyes and his skin is sallow, blotchy. Clare barely recognizes him from his photograph, but it is Tarrin Wandee.

'I was out last night,' he says. 'I rarely go out. I only went to the bar around the corner. It's more like a stall where the local boozers go. Once I start drinking, I can't stop.'

'Nor can I,' Clare says, and immediately regrets her confession.

She sees the scrutiny in his bloodshot eyes. His gaze rests on her face, trying to decode who she is from its lines and shapes, and he asks, 'Do you want a drink now?'

'No, thank you. It's a little early.'

'No such thing as too early,' he says and swallows two pills, gulps down the water. He points to a chair, which she fetches and places near him. He motions for her to sit. They face each other as the street sounds flow about them. The drone of his electric fan alternates with the rustle and flutter of papers as it executes an arc.

'Did the kids bother you downstairs?'

'No. I was fine. Do people get hassled here?'

'It can happen, if they're high on Ya Bah. Worse if they're coming down. Then they start to hallucinate monsters,' he says. 'Depends on your luck, really, and if you show them your fear. If you're not nervous and you know where you're going they leave you alone most of the time. There are so many of these kids taking the stuff nowadays in Bangkok. This neighbourhood's full of them.'

'Ah.'

'Yes. But still, it's one of the few districts left in the city that's alive. There's no life in those high rises. They're like chicken coops. Half of them are empty anyway.'

'Do you live here by choice, or circumstance?'

'You are a clever one,' Wandee says, and winks. 'By choice.'

The pills appear to take effect quickly. Wandee straightens and stretches his hands toward her, palms upward, a gesture so unexpected it disarms her. They have only just met face to face and here was an invitation to physical contact. She hesitates and reaches out, touching his warm, dry palms with hers, clammy and cold. For a moment they look at each other like two old friends who have been reunited.

'They want me to go to the hospital again,' he says. 'It's my liver. It's gone hard. It's packing up. But I know that if I go in I'll never get out. I don't want their tubes and drugs and those beeping machines reminding me that you get only so many beeps and then that's it. "I have measured out my life with coffee spoons." I have my own course of treatment ... much better than the doctors can prescribe.'

Clare's eyes dart to the table by the bed and the evidence of his pharmacopoeia. Tarrin brushes away her concerns with a movement of his hand.

'I've thought of killing myself. Sure. I don't see any reason why I shouldn't. In fact, I was thinking of doing so last night. But I got too drunk and forgot.'

They both laugh, Clare from the realization that he is possessed of a dark humour. But she, too, is reminded of T.S. Eliot: 'What is hell?/Hell is oneself,/Hell is alone, the other figures in it/Merely projections. There is nothing to escape from/And nothing to escape to. One is always alone.'

Wandee withdraws his hands and slowly reaches across the bed, and underneath the pillow. He extracts a pistol. Clare did not expect this, and her chair squeaks as she instinctively recoils. He fondles the dull metal handgun as though it were an exquisite, delicate piece of sculpture.

'I don't know if it works,' he says, tugging a little at the cocking action; Clare shields her face instinctively as the hammer moves in response. 'The guy who sold it to me assured me that it did. He even offered a money-back guarantee.' He laughs drily as he places the pistol on the table among the tobacco and the tins.

'Is there really no motivation left?' she proffers, and her mind jumps back to Eliot. 'Are you that alone?'

'I've thought about it for some years, even before I learned that I was so ill. But don't worry. The gun's to remind me of what I'm not going to do. I'm still curious, still waiting to see what's next.' He shakes his head and smiles. 'Maybe I've been waiting to meet you in the flesh. Do you think this can be true?'

Clare sees he is serious and this unsettles her again. 'I'm flattered,' she says, and, with an emphasis on exaggeration, flutters her eyelashes coquettishly.

'You know,' he says, smiling at her response. 'I don't believe in fate. Things happen for a purpose. There are no coincidences.'

'That explains things for you, maybe. But why am I here?'

'Yes. Why are you here?'

Clare laughed, remembering her mission.

'There's something I need to ask you.'

'Oh, and what's that?'

'In your book ...'

Wandee rolls his eyes and moans. 'We were having such a good time...'

Clare laughs and raises a hand to quiet him. 'In your book, you describe a young, dark-haired foreign woman in a restaurant who stops the fight between...'

Wandee looks startled, stares at her, frowns.

'Where did you get that story? Did you just make it up?'

'I like to listen to people. This town is full of stories. They're like fruits ready to be picked off a tree. Why do you want to know?'

'Because I might be that young woman you wrote about.' She tells him the story of the incident.

Wandee is intrigued. ' I got it from someone who came to stay with her cousins, just down the corridor. She came to be with her daughter who had AIDS.

'Mae Jom,' said Clare.

Tarrin nods, admiringly: 'Yes, that's right. Mae Jom. I thought her story was worth telling.'

'What happened to her?'

'Mae Jom, or the young farang woman?'

They both laughed.

'Mae Jom went back to her village. And you went back to yours, I suppose.'

He reached over to fill his glass.

'Did you really come all the way to Bangkok to ask me that when you could have sent me an email?' As he said this he stared at her with a look that demanded an honest reply.

'No, you're right. I didn't have to come all the way here to ask you that question, although I'm still convinced that it was me that Mae Jom was describing. I'd love to have met her again. But, no. It was because of something else, something that happened to me around the same time. I had to come back and make peace with it.'

Tarrin waited for her to continue. His face now wore an expression of mild expectancy like an eager child being offered a sweet. For a moment Clare debated whether she wanted to tell him everything that had weighed on her heart for so long, offer him a confession that he, as a writer could embellish and transform into a piece of fiction. But she merely looked at him and smiled.

'Let's just say that I need to confront the past before I can be liberated from it.'

He did not take his gaze off her. Suddenly and impulsively like the young, confident girl she had once been Clare reached out her hand, took his and said: 'Listen, will you go somewhere with me?'

Chapter Seven

In 1972 Clare was on the dole, living in a squat full of unshaven feminists in Brixton, thinking about the directions she might take in her life.

She held up the ticket Mickey had sent her as a birthday present, its red ink spelling out in code the flight details, and focused on the word BANGKOK. It was one of his spontaneous acts of generosity, which she knew always had some entirely selfish motive behind them. It was one of the things she liked about him.

They had met at a university exhibition of students' works. She was studying English literature, and he was at a nearby arts college. She was attracted to his funny, open face and the thick brown curls that tumbled down to his shoulders, his wiry body, his bubbly, irreverent sense of humor, his impulsive, creative nature. She was brave, he was carefree. They complemented each other. When their friendship developed into physical attraction Clare was glad to give her body to him even though she knew from the beginning that if she were to try to hold Mickey she would end up suffering. He'd already told her that he wanted no strong emotional ties or any sense of commitment, just an easy friendship;

he insisted he needed his freedom and she accepted. In fact, it suited her. She still needed to discover who she was and which direction her path lay.

Mickey was older, and graduated earlier. While his contemporaries were lured into the world of fashion photography he determined that his 'bold style' – they laughed themselves to tears when they read those words in the local paper and it became a catchphrase for a while to describe almost everything they saw – needed to be tested against a bold subject. So he ended up in Saigon at the tail end of the Vietnam War, when the big newspapers and magazines were reducing their presence in tandem with the US military withdrawal and the demand for freelancers was enough to keep a roof over his head. Clare applauded his decision, though others thought him unnecessarily reckless. They had stayed in touch; from him amusing postcards and missives from Saigon and Bangkok that never mentioned the war, from her long letters sent to a post office box number in SAIGON, REPUBLIC OF VIETNAM about the blandness of brown rice and life in a women's commune. In one of his scrawls he'd written that he had just sold a bunch of photographs, 'My best work ever', that TIME magazine was using one for its cover – '800 bucks!' – and that he deserved a little 'rest-and-recreation'. Would she join him in Bangkok, he asked, where 'the sunshine is free and I'm rich and in need of a good laugh!'

He enclosed a bank draft for one hundred pounds – 'expenses' –

and instructions on how to get herself one of the seats BOAC was setting aside for adventurous youths. Clare laughed at his audacity but complied with his instructions. She sent a postcard with her flight details.

The heat was unlike anything she had experienced before, a clammy serpent caressing her skin, squeezing the water from every pore of her body. Her ears rang with the birdsong of Thai chatter. Everywhere the faint smell of rotting flowers and dazzling colours she tried to name but gave up and simply let her eyes take them in.

She saw immediately, as they hugged at the barrier, that Mickey had changed. The boyishness was gone from his eyes, as had the long curls from his head. His hair was cut short, he'd lost weight and his skin was deeply tanned. He had acquired an air of ruggedness, enhanced by the khaki clothing and chunky boots. They talked and laughed all the way from the airport to his hotel, where they stripped each other naked and fucked frenziedly, falling back on the pillows, wet all over with sweat and exhausted with sexual satiation. They dozed briefly but soon could contain themselves no longer. They burst out laughing.

'Hello Mickey, great to see you again,' she said, realising that what she had once felt for him had not diminished.

He kissed her cheeks and her mouth and they laughed some more.

When he started talking about Vietnam, and did not stop talking until it was dark outside, Clare understood what he'd meant by needing to be with someone he trusted. It was like hearing a confession, except that he was not looking for absolution or even comprehension, just someone to hear what he had to tell; maybe he hoped that in the telling the weight he carried would be made lighter. She had followed his work when she could: his recent images from the war were leached of hope in as much as they were harsh blacks and whites or saturated extremes of colour, the red of blood, the brilliant greens of the jungle, the ochre of mud, the lifeless white of the dead.

Mickey started by telling her of the adventure and excitement and the drug-like rushes of adrenalin that accompany gunfire and the sense that one of those bullets might be looking for you. He spoke of the competitive urge to get closer and closer to the action – he learned quickly that only a short lens could give him the perspective he and the news organizations wanted.

He told her of soldiers dying around him, not quickly from one clean shot but slowly, and screaming from gaping wounds. Civilians, whose only loyalties lay with their ancestral lands, died more quietly. They were lined up and gunned down, or silently slaughtered from above.

Mickey said the smell of the spiky durian fruit, which had repulsed him in the markets, was close to the odour of death. He had thought himself to be neutral, a Brit observing and recording a foreigner's war, but the barbarous inhumanity he had observed through his viewfinder had challenged his courage. Not all the people in his photographs wanted to be there within the frame but some, more than a few, were thrilled by the killing; those faces scared him the most.

Clare listened. That was all Mickey wanted. She would fetch him cold beer as he talked, emptied ashtrays, adjusted the shutters to catch any passing breeze, switched on lamps as the light faded into the night. His easy humour had not totally disappeared, but his remarks were less prone to flippancy and now had a morbid edge washed with cynicism, much in the way he had once shown her how bathing a portrait in sepia could change entirely how the viewer perceived an image.

They had at some point migrated to the balcony and Mickey, perhaps exhausted by his effort to purge himself of the sins of others, fell silent and let the nighttime noise of Bangkok swallow them up. Clare sat and waited. Maybe it was a minute, maybe thirty. Time had lost meaning, until Mickey coughed wetly and spat into a pot and announced they were going to a Chinese restaurant for dinner with some of his – and he wiggled his fingers to make air quotes – 'colleagues'.

'Your fellow misfits, you mean,' she said, and he laughed and she was glad he still could.

'Yeah, you could say that. They're all mad, completely bonkers. Anyone who's got anything to do with this war has to be.'

At a sticky table littered with empty bottles and stale ashtrays Clare was introduced to three men, one a French photographer whose attempt to look rakish with an earring and bandanna tied damply around his neck was spoiled by his unbuttoned fly; a local Thai journalist, who bowed, palms together in formal greeting and whose bright, white smile dazzled her; the third a middle-aged, ruddy faced and moustachioed Englishman in a bow tie whom Clare gathered did something or other with Reuters and whose booming voice and encyclopedic knowledge of Bangkok nightlife had them deferring to him as 'The Captain'. Clare noticed that when they were introduced a look of complicity passed between him and Mickey.

They had moved on from beer and were drinking the cheap local whisky called Mekong, like the river, mixed with ice and soda, and plenty of lime juice to block its unpleasant, perfumed taste. One empty bottle was replaced by a full one, which The Captain cracked to mix their drinks. 'Now that we have the company of a delicious lady,' he bellowed, 'it's time for some delicious food.' And he summoned to his side a thin, dark-skinned girl with a long ponytail who nodded as the big Englishman launched into Thai, his accent so light and delicate Clare had to laugh, the others supporting her observation of what was clearly an in-joke.

She felt immediately among friends. The girl stared intensely at Clare with a frank, innocent but quizzical look that transcended language: why was she there, a woman, alone in male company, and drinking as hard as the men?

Clare was instructed by The Captain to eat everything in front of her. 'Don't think about it or you'll starve.'

They were halfway through this banquet, during which the conversation roared along like a rollercoaster, with surprise twists and turns and raucous laughter as it dipped and climbed, when a group of American soldiers settled into the next table. Clare's attention was drawn to their loud Hawaiian shirts, their ill-fitting chinos and close-crop haircuts; young men with clean features and ready smiles, but with eyes that looked hard and edgy and seemed focused on something beyond the sightline. A few were unsteady on their feet and had to be helped into their chairs.

The Captain shouted a greeting and one of the soldiers waved back.

'And how goes the war?

An innocent enough question, Clare thought, but she felt a jolt of tension cut through the air as their little rollercoaster came to an abrupt stop. Her dining companions became suddenly alert. One of the soldiers turned and crossed the short space separating the tables, pointed a finger in The Captain's face – a gross insult in Thailand, Clare had already been warned, and one noticed by the patrons of the restaurant, where there was a sudden hush, and chairs began to shuffle.

'We're winning, man,' the soldier said. 'You get it? We're going to kill every fucking gook over there before we're through.' His stare was icy, his body language aggressive and menacing.

The Captain did not move, but held the soldier's eyes without blinking.

One of the other soldiers muttered: 'Cool it, man,' and stood up to draw him back to their table.

That might have been the end of it, had Mickey, now very drunk, not blurted out aggressively, 'You're fucking finished, mate. They're kicking your arse. Accept it. You've lost. The war is over. There's no more to do except murder them, so just fuck off and go home.'

The young soldier turned, every hard muscle taut, his jugular pulsing fast. His companions stiffened, at the ready. As he took a step forward he reached behind him and from under his flowery shirt pulled out a handgun. There was an audible silence in the whole room. For a second no one moved.

'For Chrissake!' screamed Clare, springing up from her seat like a tigress and now standing between the two tables. 'It's my birthday and I don't want it fucking ruined by some bullshit scene. Sit down and shut up. The lot of you.' She was shouting this straight at the young man carrying the weapon.

In a heartbeat, it was over. The anger drained out of the soldier's face. He put the gun back in its holster as he looked at her, blinked as if coming out of a trance, smiled and, after a few moments, began laughing. 'I'm sorry, Ma'am. I don't want to spoil your birthday.' He bowed and stepped back.

The other soldiers raised their glasses and chorused: 'Happy Birthday!'

One of them shouted, 'You're one ballsy chick.'

Crockery clattered and chatter resumed at the tables around them and, for the moment, all seemed right with the world. It can't have been much later that the manager was whispering in The Captain's ear. He declared shortly thereafter that their meeting was adjourned.

'Don't know what came over me,' said Mickey, sprawled across the bed, his face buried in a pillow. 'You did great.' He had not spoken since they'd left the restaurant, and had waved off the others' suggestion that the night was still young.

'I'm glad nothing happened,' Clare said. She was trying to understand, but Mickey wasn't making things easy. 'Those soldiers ... they're just pawns in someone else's game.'

He lurched into a sitting position, facing her. 'You could not be more wrong ...' He was fumbling his words. 'Those soldiers ... they were Special Forces. You saw. They like their guns. They do the dirty work, cleaning up messes, making sure there are no loose ends. No one makes them do anything. They're all volunteers.'

Mickey went quiet and there was the same anguished look from earlier.

'How can you bear it?' she asked finally.

'I can't ... I hate it all ... but ... I'm hooked.'

Mickey slept heavily until early afternoon and after a long shower was back to his bouncy self. He announced that he had a surprise for her. A taxi took them out of the city and into a landscape of flat, green rice fields bordered by dirt roads. Mickey had the driver turned down a track and they pulled up in front of a group of wood and palm-thatch shacks, around which were parked battered and grubby pickups and sedans.

Mickey said something in Thai to the taxi driver and handed over some bills. They got out, and the driver pulled away from the shacks and stopped again a short way back up the track, switched off his engine and opened a newspaper.

Clare could smell the sweet scent of burnt marijuana and frowned. In London they had both tut-tutted their disapproval of drugs and disdained those who'd used them; for her, it was the result of being drilled in the suburban hypocrisy that drugs were criminal and a short path to hell (whereas alcohol was perfectly acceptable); for him, it was a reluctance to lose touch with reality. Mickey just smiled and led her inside, where the light was so dim Clare was blinded until her eyes adjusted. She heard Mickey's name spoken and could make out body shapes as he led her gently to sit on a mat woven from what her fingers worked out was some fibrous material. The thin face of the man next to her came into focus. He was almost gaunt, with thick curly hair and a direct gaze that looked wild and dangerous. He was wiry, wearing only a pair of faded short canvas trousers; his upper body was covered with tattoos, strange looking designs and letters on his chest and his back, and running up his arms near-human, vaguely mythological shapes. She started to smile an apology for looking at the peculiar body art, but the man seemed not to mind and began twisting his torso this way and that to better show his more elaborate inking. He looked at Mickey and rubbed his two index fingers together and laughed softly to himself.

Clare saw there were people lying on mats all over the room, which was wide and bare, some on their sides and propped up by pillows, others on their backs looking up at the high vaulted roof. They were all Thai, except for a young man with long blond hair and a short, well-trimmed beard leaning against the wall. He nodded to Clare when their eyes met and raised one hand to greet her.

The tattooed man pulled some herbs from a sack by his side, deftly separated the seeds onto a plate and chopped and scraped up the leaves on a thick wooden block, its surface a dark patina of oil. He looked like a sous-chef preparing parsley or coriander, humming a chant-like melody, and when he had finished he put the knife by his side and held the bamboo pipe in both hands above his bowed head. Clare thought it all very spiritual, as if he were communing with gods. She looked at Mickey, but he was watching the pipe, inlaid with a silver dragon, and the bowl finely carved to look like the head of a flower. The tattooed man filled the bowl with a pinch of the herb and poured water from a kettle into the pipe itself before handing it to Clare.

She shrank back and held both her palms out to refuse, but the man kept smiling and nodding until she took it from him. He lit the bowl, but her draw was too weak and the flame didn't catch. Still smiling, he took the pipe, lit it, and sucked hard until the bowl emptied with a small pop. His eyes remained on Clare as he slowly exhaled, the smoke shrouding him in a blue haze. He prepared the pipe again and offered it to Clare and this time she got some smoke, which made her cough and splutter until her eyes were running with tears.

'Buddha grass,' the man said. 'The Americans call it Buddha grass.'

Mickey took his turn before each of the others, shuffling to the middle of the room to take a hit before returning to their prone positions. No one talked; Clare heard only birdsong and the breeze playing with wind chimes, the humming chant of the tattooed man and the tick of his sharp-edged knife as he cut and prepared more grass. He would from time to time deliver a lyrical monologue in Thai and let out a throaty laugh. Some of the others laughed too. Even Clare laughed, though she had no idea what he was saying. The young blond foreigner took a guitar out of a case and started playing. It was a slow folk/blues number. It sounded familiar to Clare as she lay on her back on the floor, closing her eyes and listening to the resonant chords and delicately picked strings of notes. She imagined herself floating on a long wide expanse of water towards a horizon she never reached.

When Clare opened her eyes, the musician was gone and a young girl in a sarong and cotton top was serving tea and a bowl of brightly coloured fruit was being passed around. Clare sat up when the girl brought the bowl to her and she selected something red, shaped like a pear, its texture soft and cool to the touch. She bit into it and thought it the most wonderful fruit she had ever eaten in her life.

'Chompoo,' said the tattooed man. 'Good to eat after smoking.'

It was dark when they made their way back to the city. Mickey stared out of the open window, a smile on his face, content. He told the driver to drop them in front of a brightly lit market by the river. There were flowers everywhere. The colours and scents made Clare's head reel and her blouse was sweat-soaked from the heat and stuck to her back. Mickey was holding her hand and manoeuvring her through the crowd; he seemed to know where they were going and Clare let herself to be led along narrow pathways weaving between the stalls.

She was still high and disorientated and suddenly felt panicked. She squeezed Mickey's hand. He understood, they stopped and he put an arm around her shoulders, whispered 'shhhh' softly in her ear.

A Thai folksong blared from a pair of speakers suspended above the market, its sinuous rhythm perfectly in sync with the slow-motion movement of the crowd. Clare let her eyes wander, kept touching flowers, delicately, and stroking the fruits and vegetables, like a child venturing out into the material world for the first time. Wonderment. And when she looked up there were laughing eyes staring into hers as though they understood precisely what she was experiencing.

Mickey led her on through the market, slowly, but as they reached its far edge she became frightened again, as though everything around her – the intense activity, the bright lights, the unfamiliar colours and smells, the foreignness of the faces, the sing-song speech, everything, including her own body – melted into a kind of whiteness. It lasted just a few seconds, but she felt as if she had been transported to a place where substance ceased to exist.

Their lovemaking was slow and sensual that night. After, as they lay sweating side by side, Clare felt she was in a state of grace.

'If I'd known it was like this, I'd have started smoking years ago,' she said, her laugh cracking into a slight cough. 'Now I want to try it all ... acid, opium ...'

'Don't go down that road,' said Mickey, lighting a cigarette; Clare watched the ember at its tip flare. 'You'll end up handing flowers to policemen, or joining some sect and chanting mantras in Trafalgar Square. Stick to your gin and tonic, a good and thoroughly British drug.'

'I had an odd moment in the market,' she said, and began to describe the experience before Mickey cut her off.

'It's just an illusion ... the grass does funny things.' He grew serious. 'I mean it, don't go there. You're vulnerable and you're too straight. You'll end up taking stronger stuff and you won't be able to get out. It's a waste of time. There's so much to do in this world.'

'So why are you into it now?'

'To escape,' he said. 'To block things out. To keep the demons at the gate. I have to. Otherwise I'd go crazy with what I've seen. I need it just to get me to the next day.'

Mickey took her back to the shack one more time before his ten days were up. The tattooed man wasn't there. 'Away on business,' they were told. Clare didn't feel the same magic as they smoked, though the grass was just as potent. As they were leaving, Mickey was given two cartons of cigarettes.

'Can't you get cigarettes in Vietnam?' she asked him as the taxi drove them back towards the city.

Mickey smiled.

'Not these. They're special.'

'Oh ... I see.'

'No, you don't. This stuff is why the Americans have lost the war. A few puffs and it's like: "What war?" '

'You really mean this, don't you?'

He nodded. 'Who wants to fight and kill when they've been to the other side of reality?'

They were silent for the remainder of the ride. That night Mickey asked her to stay on in Asia, with him; just like that, without warning. It was typical of him and it made her laugh because she thought he was joking.

'Seriously. I need you to be with me. I can't be here on my own. Not any more. It's getting too weird.'

This new fragility confused her. Could she trust him, she wondered.

'Listen, I'll marry you, if that's what it takes to have you back here.'

Clare was taken by surprise at this casual, flippant, almost rhetorical proposal. It made her nervous. She never expected this from Mickey. At the same time she felt elated.

'But... I can't just drop everything in London,' she said.

'So go and sort it out, and then come back.'

It was their last night together and it was all arranged. She would be two weeks in London – time to take her money from the post office savings account, to say goodbye properly to her parents and her sister. They would ask her many questions and she would tell them she did not know how long it was going to be before she saw them again. She was off on her adventure in Asia, and she wanted to be with Mickey whatever the future held. She had always been in love with him and now he wanted her by his side.

They'd gone to a restaurant in Chinatown and got through a bottle of Mekhong. 'Can I have a decent drink now?' Clare had said after the meal. 'Take me somewhere exotic. I need cheering up. I don't want to leave you tomorrow. I'm a bit sad.'

They had walked to the river hand in hand.

'This place has the best view in Bangkok,' Mickey said as he led her through the doorway of a small, grubby old hotel whose name, 'The Chao Phya Inn' was displayed in garish green neon. The lobby was empty except for an old Chinese woman dressed in black sitting at the reception fanning herself who nodded a greeting as they made their way to the open terrace at the back that jutted out over the water; rusty tin tables and chairs, fairy lights strung across the metal uprights, the tugs gliding by in the dark water, the tall pointed stupa of a temple silhouetted opposite. How can she ever forget?

They were the only customers and for a while they sat in silence enjoying the view in front of them. Clare is glad Mickey has brought her to that romantic spot. She could understand what it was that attracted him to that city.

'I'll have a beer and a chaser, if ever we get served,' she said as she leaned over to kiss him. Then she got up from the table and headed for the washroom, passing a sleepy-looking waiter who had appeared out of nowhere as she did so.

Barely ten minutes later, on her way back to the terrace she passed him again in the lobby and this time he looked at her with a curious smirk on his face. From the direction of the river she heard laughter and the booming voice of 'The Captain' speaking Thai. Instinctively she stopped at the door. They were all facing the river so that at first they could not see that she was there. But then in unison they turned towards her.

In the theatrical light The Captain's face was flushed to a deep purple like a drunken figure in a pantomime. Beside him, with one hand lazily draped round his shoulder stood a young boy who could hardly have been in his teens. He had dark skin and long hair, cut like a girl's. He was wearing a T-shirt, tight jeans and white sneakers. Next to the Englishman he looked tiny. Another boy, dressed in the same fashion, was standing slightly apart smoking a cigarette. He smiled at Clare.

The third boy was giggling as he struggled to pull himself up from Mickey's lap. One hand was round Mickey's neck, the other on his crotch. It was obvious they knew one another. For a moment Clare stood frozen in shock as she took it all in. Mickey, pinned down in his seat, did not, could not, move at first but merely stared at her with pleading eyes. Just before turning away Clare saw The Captain raising his hand before shouting something out. She thought she caught the words, "our lovely boys", but she did not wait to hear any more as she hurried from the terrace.

In the heavy silence of the hotel room Mickey had lit up one of his special cigarettes.

'Do you need that before you tell me what the hell was going on out there?' Clare's voice was trembling, not just from anger, but from the fear of what she was about to learn.

'I didn't know they were going to turn up,' he said blowing out the smoke that smelled faintly of burning plastic. 'Bad timing.'

'What can I say, Mickey? I'd like to think that it's being around the war that's done this to you. But it's more complicated than that, isn't it?'

Mickey nods, without turning to look at her.

'What was all that talk about needing me by your side? You proposed, Mickey! Why? Am I your cover? Do you need a wife to disguise yourself?

Pretend to the world you're something you're not?'

Mickey buried his head in his hands. Clare's first instinct was to reach over to stroke it, to soothe him. But the imprint of what she had witnessed gave her pause. She knew she would never touch him, or be touched by him ever again.

She had not allowed him to accompany her to the airport. Six months later, Clare received a letter from Mickey's father. The British consul in Chiang Mai had informed him that Mickey had been killed in Laos, near the Thai border. The circumstances of his death were unclear. He had been accompanying a tribe of Hmong mercenaries when they were caught in a firefight. Mickey's body was found by a Thai border patrol. His father said he had spoken affectionately of her, and he thought she should know the tragic news. There were details of a memorial service. She never replied.

Chapter Eight

Nai Pot – for that was now the identity that she had given him in her mind – is leaning against the car, chatting with some of the youths from before. They are laughing easily together and when they see Clare emerge they cheer for her. She adjusts her sunglasses as they jump around her and then tussel to open the rear door of the limo. They pay little attention to Tarrin, whom they know but who remains for them a weird and unpredictable outsider, someone to be left alone. Nai Pot says something and they disperse, waving, still laughing, and Clare and Tarrin climb inside the cocoon of cold air. Nai Pot says nothing as he steers the car back the way they came.

'Can you take us to Chinatown?' she asks.

As they cross back over the river the swollen clouds hanging over the city erupt with lightning and thunder and a furious downpour that defeats the windshield wipers. The car is forced to a halt. The rapid tympani of rain reverberates on rooftop; it is cacophonous, yet hypnotic. As it abates, Nai Pot makes some adjustments to dials and buttons on the dashboard and, with a deep breath, turns the limousine quickly onto the main road.

Twenty minutes later when they crawl into Yaowaraat, the road that slices through Chinatown.

Nai Pot turns briefly to Clare who is sitting lost in thought.

'Anywhere in particular, Madam?' he asks.

'Oh please don't call me "madam",' she says.' I don't run a brothel. 'Do you know the Chao Phya Inn? Does it still exist?'

Five minutes later they park in front of a building nothing like the one she remembers. It is a hotel and the name is the same but it has been totally remade into one of the 'boutique' places that have sprung up all over the city in response to the wealthier backpackers. None of the shabbiness remains. The lobby is unrecognisable; low rattan furniture and carefully tended potted plants, an old-fashioned ceiling fan that whirs overhead. The girl at the reception is smartly dressed. She looks confused and slightly dismissive, unsure if the odd couple in front of her – a grey haired farang lady in a flowery dress, and a younger, unshaven Thai man in scruffy jeans and flip flops – carrying no luggage, have come for a short amorous stay. In the end her face melts and she puts on her welcome smile, and is about to speak. But Clare puts up her hand and then points to the French windows where once a rickety door had stood. They are closed and the canvas awning has been fully stretched out to protect the space from the recent rainstorm.

She stands looking at the empty terrace and the elegant tables and chairs and the terracotta tiles, all damp and shiny, and the river beyond the low wall, choppy and brown, with the boats bobbing in it and the temple stupa that she has never seen in the daylight rising up to the grey, threatening sky. She remembers everything as if time did not fly but stood still and she knows now, beyond any doubt that her heart atrophied that night, and never recovered. *It explains nothing,* she says to herself, *and everything.*

Suddenly the skies break open once more and in that moment she feels a hand gently on her shoulder, a ghostly touch that startles her out of her memories.

'Are you OK ?' Tarrin's gentle voice is hardly audible against the sound of the rising wind and the droplets of rain hitting the metal tabletops, against the noise of the riverboat engines.

'It's going to pour down again. We'd better go.'

Clare nods before moving away.

Many of the small lanes leading from the main road are flooding and impassable. There are lines of people wading knee deep through muddy water. It is a sight reminiscent of Tarrin's novel. Nai Pot slows the car down and with this Clare sits back in the soft leather seat and closes her eyes, lulled by the music of the rain.

Meeting Tarrin, and then their visit to the terrace by the river have exhausted her. But she knows she was right to come back to Bangkok. In a piece of fiction she might have seen it as a journey to reunite with the brave young woman from the restaurant, who could stop a fight. But it was not true. Life is not fiction. It was too late to recover that young woman and excuse her for not daring to go on questioning the social norms, for having chosen a quiet, secure life fostering the words of others rather than being a writer herself. And yet she knew that a certain quality of her heart had been revived. Her return to Bangkok was to look plainly at the pain attached to the memory of the city – something she has carried deep within her without ever sharing it with anyone, not with her sister or her lovers. The horror of her discovery of Mickey's dark side had not been dulled by the years. It has always been a part of her. It was the decisive moment of shock that took away her trust.

Returning to the physical place where, in a sense her heart had stopped, was a pilgrimage to a part of herself that she had thought she'd lost. 'Forgiving is not forgetting' – it was a phrase someone had once told her and which had not meant much till that afternoon. Now, after all these years, just before she would forget everything, she knew that she was free to forgive because she was no longer afraid of facing either the past or the future.

She was grateful that Tarrin had been there as a witness. She had not explained anything to him when they had driven back. There was no need. She knew he had sensed that something important and transformative had happened. She was amused by the thought that he might later have made use of the incident, elaborated on it in his fiction. Towards the end of the journey they'd held hands once more. It was a natural, innocent gesture of friendship that elicited in her a feeling of warmth she had not known for a long time. Before they parted she told him: 'I'm alive now, again. And I'm ready to take my journey.'

'Journey? Where to?'

Clare stayed silent for a while as she looked out at the soaked cityscape: the plastic awnings sagging heavy with water, the people rushing from one sheltered doorway to another, all the vehicles seeming to glide on a surface that reflected the neon-lit signs and the trembling contours of the tall buildings. Everything looked alien, dreamlike.

'A journey into otherness,' whispered Clare finally as she turned to him. 'Through the curtain of rain.'

Tarrin laughed. 'Yes ... I like that,' he said. 'Otherness.'

She opens her eyes and Nai Pot is facing back towards her. She does not know how long she has been asleep, nor where she is. For a second she merely blinks at him in her confusion. He smiles at her patiently.

'We are at hotel now.'

The storm has gone, the sky is bright, and he opens the door for her. She climbs out of the limousine, touches his hand, smiles.

'It was a bad one. Were you frightened?' he asks.

'Yes, I was,' she says. 'But I'm not now.'

Chapter Nine

An Ending

When the end came Clare was glad they'd made no fuss. A leisurely lunch in a chic restaurant near the office, with the CEO and Paul presiding; the four-volume Blackwood edition of *Middlemarch*, beautifully bound and also very expensive, for her years of service; and a cheque – altogether a generous farewell, much to her liking.

Cutbacks, Paul had explained, crisis in the publishing industry, uncertainty over eBooks, seepage of advertising to the web, and so on. To save jobs lower down, and as her official retirement was not so far away, the directors hoped she would understand why she was being let go. She had to smile at that particular euphemism: she would be letting go of a lot of things soon.

Earlier in the week Paul called her into his office, closed the door, and told her that Tarrin Wandee was dead. He was found on his bed by his neighbours. It looked like a peaceful, natural death.

He shook his head, perplexed. 'What a pity. Just when he was on the point of international recognition...You were probably one of the last people he spoke to. You were so lucky to miss those terrible floods.'

'Yes.' Clare did not outwardly show any signs of shock. A part of her had been expecting to hear about him, but she had not expected this. 'I'm very sorry to hear it,' she added. 'He was...'

'This was sent to us,' said Paul without waiting for her to finish. He pushed the large manila envelope across the desk towards her. 'Tarrin's last notes. They're interesting. He mentions you.' Paul's face took on a serious expression. 'Seems that what happened between you two really affected him. What exactly <u>did</u> happen between you?' The look that accompanied this remark was new to her, which was surprising given how long they had known one another. It was as if he were expecting to hear some salacious piece of gossip. She had not read the notes, so she could only guess at what Tarrin had said. She gave him a silent shrug in reply.

Seeing that she could not be persuaded to yield up any details Paul continued: 'But what should we do now?' he asked, opening both his palms up to the ceiling in a gesture of supplication. He did that whenever he had already made up his mind but wanted to appear to be in consultation with others. Clare detected something tricky in his tone, and she didn't like it. She had written her short, introductory piece on Tarrin and had presumed the project would go ahead as planned.

'What are you talking about?'

'I mean, is it really worth the investment? We won't get much mileage out of him now, will we?'

Clare glared across the wide table at her colleague. She understood precisely what Paul was thinking. In all the time they had worked together, she had never once raised her voice to him, or to anyone else. She had never dared. Now she could feel the anger rising in her.

'No, Paul, you can't do this. I won't let you. Contracts have been signed. The pages are laid out. Plans were made, even if they can't now be pursued.'

Paul was taken aback, surprised by her reaction and by the command in her tone. He had never known her to be like this before in all their years of collaboration.

'I know how you must feel, Clare. You were his editor. But, really...'

'No you don't. I warn you, Paul.' With those words Clare rose from the chair and left the room.

Paul summoned her again the next day to tell her, as if he had planned it all along, that the story would go ahead. The directors wanted Clare to write a preface.

She was flattered and tempted to refuse – knowing how difficult it would be for her both emotionally and now physically, with the knife of Alzheimer's hanging over her head. The trick, she thought, would be to write it quickly, and as soon as possible. It would be her last assignment before she packed up her belongings. Apart from that, her work was done.

The initial prognosis was off the mark and the Alzheimer's advance gathered pace. Within six months she had trouble remembering the layout of her own apartment. London, where she had grown up and lived all her life, might as well have been Tokyo.

Clare, when still lucid, had been so thorough in her preparations, so calm and matter-of-fact that her sister had asked, 'You seem so well, so at ease with everything. But are you really OK? Are you afraid?'

'It's funny,' said Clare. 'Someone in Bangkok asked me the same thing.'